Understanding **LED**
Illumination

Understanding **LED**
Illumination

M. Nisa Khan

CRC Press
Taylor & Francis Group
Boca Raton London New York

CRC Press is an imprint of the
Taylor & Francis Group, an **informa** business

CRC Press
Taylor & Francis Group
6000 Broken Sound Parkway NW, Suite 300
Boca Raton, FL 33487-2742

First issued in paperback 2019

ISBN-13: 978-0-4665-0772-2 (hbk)
ISBN-13: 978-0-367-86710-2 (pbk)

Library of Congress Cataloging-in-Publication Data

Khan, M. Nisa.
  Understanding LED illumination / Nisa Khan.
  pages cm.
  "A CRC title, part of the Taylor & Francis imprint, a member of the Taylor & Francis Group, the academic division of T&F Informa plc."
  Includes bibliographical references and index.
  ISBN 978-1-4665-0772-2
  1. Light emitting diodes. 2. Lighting. I. Title.

TK7871.89.L53K53 2013
621.32--dc23   2013009380

Visit the Taylor & Francis Web site at
http://www.taylorandfrancis.com

and the CRC Press Web site at
http://www.crcpress.com

To the memory of my father,

Mujibur Rahman Khan

Contents

7. LED Replacements for Incandescent and Linear-Tubular Fluorescent Lamps   193

Preface

The field of lighting has become very dynamic in recent years, primarily due to the fast and notable improvements of light-emitting diode (LED) lamps. The rapid development is a result of many dedicated scientists, engineers, and academics who recognize the potential of significant energy savings in global lighting consumption. LED lamps are also drawing attention from many lighting designers and the entertainment industry because they can be produced to emit white light as well as many single colors to fill the visible spectrum. Their light characteristics can be electronically controlled to produce a variety of desired lighting effects and full-color video images. Solid-state electronics have transformed the world due to its versatility, efficiency, and large-scale manufacturability. Can solid-state lighting also transform the lighting industry? What would be required for LED lighting to become mainstream?

Rather than drawing parallels to electronics with regard to future prospects, a thorough investigation would be more beneficial to address these questions. Full understanding of design, specifications, and quantifications of LED lamps and luminaires for various lighting applications is essential in order to manufacture solid-state lighting products to help save energy and beautify our environment. The author's intention is to provide such broad comprehension while being mindful of the aesthetic nature of lighting that uplifts emotions. The book is intended to take the mysteries out of solid-state lighting for lighting designers *and* to help LED scientists and engineers effectively design their products to provide high-quality illumination.

Most LED scientists and engineers are duly engaged in substantial amounts of rigorous work on semiconductor physics, materials, and optoelectronic device engineering. Understandably, it is atypical for many such professionals to concentrate fully on all aspects of lighting. However, an LED engineer pursuing a final luminaire still needs to incorporate all the proper considerations from the lighting perspective. Thus, it is important for LED lamp and luminaire developers to understand and appreciate lighting fundamentals and illumination criteria. In the lighting industry, while lighting scientists and designers understand and appreciate illumination better than traditional LED engineers and scientists, they are not familiar with the intricacies of LED science and technology. Therefore, they are, by and large, unable to contribute toward LED lamp development for general lighting. In order to help both communities, many definitions and descriptions are kept rather basic and restricted to give the reader an easy but, nevertheless, a fairly wide understanding of light and LED illumination science. The author attempts to describe comprehensively many advantages, disadvantages, and bottlenecks of LED lighting technologies.

The solid-state lighting industry has made remarkable progress in the last few years, primarily in terms of improving lamp efficacy and color quality to rival and even outperform fluorescent technologies in several applications. The areas that still need more innovation are enhancing scalability and light distribution properties. The book addresses these needs by focusing on light propagation and distribution characteristics of LED light sources for general illumination applications. It draws on all the lighting industry's adopted terms that describe lighting for designers to create and for users to appreciate desirable illumination in living spaces. It describes current methods as well as the author's own method of generating broad and diffused light from LED lamps. In order to be considered as a practical replacement lamp for general lighting applications, it is essential for LED luminaires to utilize secondary optics methods such as diffractive or integrated optics to distribute and scale light that can provide the same desirable large-space illumination produced by many incumbent lamps.

The first three chapters of the book address lighting fundamentals and technologies, followed by LED science and technologies—first at the device level and then at the module and luminaire levels. Chapter 4 provides a comprehensive portrayal of lamp measurements and characterization in terms of standard photometry and colorimetry. Chapter 5 addresses LED lamp designs and suitability for various applications amid the current challenges. Chapter 6 discusses lighting theory and simulation techniques to differentiate LEDs' illumination behavior from other lamps; these are then applied to a novel design for an omnidirectional LED lamp. In the final chapter, several LED replacement lamps for household and commercial lighting uses are characterized, and a novel design for improving tubular LED replacements is discussed. As a general theme, the book tries to clarify some frequent misunderstandings of LED light sources—something that is much needed in the lighting industry.

The author hopes that this book will be uniquely beneficial to LED and lighting industry professionals and educators. In particular, the book is intended for scientists, engineers, and technical people from all disciplines developing LED lighting products. It is intended for engineering faculty members, graduate students in engineering, advanced undergraduates, and working engineers and scientists with a need to understand the fundamentals of illumination and how to apply such understanding to design LED lamps.

Scientific and technical people from disciplines other than optics and lighting who are genuinely interested in learning about illumination, energy efficiency, and reducing electrical power consumption may also find the book beneficial. In addition, the author hopes that a general audience with some technical background and an interest in lighting will find this book "illuminating."

Acknowledgments

I am indebted to many people who contributed to this book in a variety of ways. I am grateful to the people and institutions that helped me develop and grow as a scientist and engineer. These include my alma maters—Macalester College and the University of Minnesota—where I learned valuable essentials in mathematics, science, and engineering. A great deal of tribute goes to Honeywell, where I learned much about advanced solid-state technologies and where even the senior technical staff members never made me feel insignificant despite my young age. I would like to express my deep gratitude to Professor Marshall I. Nathan, for being my beacon at the University of Minnesota and for sharing the extraordinary experience he had in the 1962 semiconductor laser race, during which the first visible laser diode and LED by Dr. Nick Holonyak were discovered. I am indebted to the many former colleagues from Bell Labs, in particular Dr. Charles A. Burrus, who prepared by hand literally millions of semiconductor diodes and LED samples for many scientists, including me, to investigate.

My appreciation goes to the annual event, LIGHTFAIR International (LFI), where I learned valuable material on lighting from the courses offered by the LFI Institute supported by the Illuminating Engineering Society (IES) and the International Association of Lighting Designers (IALD). Concerning the field of lighting, my special gratitude goes to John Williams, president at YESCO (Young Electric Sign Company), for his support and encouragement. His meticulous approach to solving lighting and illumination problems during our collaboration catapulted my interests toward the lighting aspects of LED-based lamps. Lastly, I would like earnestly to thank Luna Han of Taylor & Francis for her effective guidance and encouragement in my writing of this book.

About the Author

M. Nisa Khan received her bachelor's degree in physics and mathematics from Macalester College, St. Paul, Minnesota, and her master's and PhD degrees in electrical engineering from the University of Minnesota, Minneapolis. During her studies, she worked as a research associate for 9 years at Honeywell Solid State Research Center in Bloomington, Minnesota. After completing her doctorate, she became a member of the technical staff at AT&T Bell Laboratories (now Alcatel Lucent) in Holmdel, New Jersey, and spent most of her 6 years at the Photonics Research Laboratory at Crawford Hill conducting pioneering work on 40-Gb/s optoelectronic and integrated photonic devices. Dr. Khan then worked on optical communication subsystems at several other companies, including her own venture-backed start-up companies in New Jersey. In 2006, she started an independent research and engineering company in LED lighting and has since been involved in innovation and technology development for making solid-state lighting more suitable for general lighting. As an independent consultant, Dr. Khan performs feasibility studies for LED lighting used in entertainment and signage industries and offers platform design and development solutions for general lighting applications. Since 2007, she has been writing the "LED Update" column for *Signs of the Times* magazine, which has been serving the electric signage industry since 1906.

① Introduction

1.1 Introduction

Light is a vitally important physical resource for all living beings. Aside from providing us with vision, light is inherently connected to all life, which we describe as photobiological phenomena. While human beings have always had this integral relationship with light, our significant understanding of its properties and behavior only started a few centuries ago, after which the major discoveries unfurled in steps. Starting in the late 1660s, Sir Isaac Newton initiated the corpuscular theory of light, explaining that light was made up of little particles, or "corpuscles" and that each of these particles did not have the unique color of "white," but rather these white light particles were composed of a spectrum of discrete colors that can be separated with a prism [1]. In practice, many optical phenomena can be handled with Newton's theory, which forms the basis of geometric optics or ray optics. Around the same time, Newton's adversary Robert Hooke deduced that light did not have behaviors of a particle, but rather that of a wave, from which Christian Huygens developed his wave theory in 1690 [2]. However, the wave theory was not vindicated until Thomas Young and Augustin-Jean Fresnel did interference experiments that proved that light has a wave-like property that could not be supported by Newton's corpuscular theory. Subsequently, diffraction theory was established and the study of physical optics, otherwise known as wave optics, was set forth [3].

Wave optics was further embraced when it was unified with electromagnetic theory by James Clerk Maxwell in the 1860s, establishing that light waves were in fact electromagnetic radiation [4]. At the turn of the twentieth century, Max Planck and Albert Einstein formulated the astonishing theory revealing that light has both wave-like and particle-like properties, which they explained using quantum mechanics [5,6]. When viewed as a particle, light was then referred to as "photon," falling in the category of "boson," which behaved based on Bose–Einstein statistics, in contrast to electron particles, which are "fermions" following Fermi–Dirac statistics [7]. Based on subsequent developments of quantum mechanics and electrodynamics, the twentieth century became an energetic and exciting field for physics—optics in particular—leading to amazing discoveries in astronomy and various engineering fields including light-emitting diodes (LEDs), lasers, photodetectors, fiber optics, and others.

From the turn of the twentieth century, as the understanding of light at the fundamental level progressively developed in the field of optics as a branch of physics, lighting development for illumination applications took an unprecedented turn at the same time, starting with the invention of the *practical* incandescent light bulb generally credited to Thomas A. Edison (albeit with much controversy). The significance of the tungsten light bulb was not only its mere invention, but also how quickly it became ubiquitous in households, particularly in the United States, because of its successful deployment via common electrical systems, affordability, and practical benefits. From 1914 to 1945, lamp sales went from 88.5 million to 795 million units, reaching more than five lamps per person per year [8]. Lighting science and engineering became a field in its own right, primarily for illumination and human vision applications, benefits of which have always been imminent and extraordinary.

Although incandescent lamps are widely used in households and commercial buildings because of their practicability, they consume a great deal of electric energy because the incandescence process typically converts only a few percent of electric energy to visible light and over 90% to invisible thermal radiation. As other lighting technologies have become practical and more energy efficient, incandescent lamps have started to be replaced gradually in many applications. These include such electric lights as linear and compact fluorescent lamps, high-intensity discharge lamps, and LEDs.

Enormous improvements in LED lighting technologies in the past decade have led many to wonder if LED lighting will be chosen for nearly all illumination purposes. The idea is becoming increasingly popular as LEDs' theoretical luminous efficacy, at the small-scale source level, is calculated to be about twice that of the current state-of-the-art fluorescent lamps. However, the question remains whether this higher efficacy can be scaled for practical-sized lamps that can illuminate omnidirectionally. This book investigates such challenges of LED lighting and analyzes some solutions for practical LED lamps. Recognizing these challenges and further developing various existing solutions and perhaps adding new solutions, LED lamps for general illumination applications could take a significant step forward.

1.2 Lighting Fundamentals

1.2.1 A Very Brief History of the Study of Light

According to modern history, investigation of light, vision, and color began when the Greek philosophers Plato, Aristotle, Democritus, and others provided early philosophical and psychological descriptions of light and color. The development of light science or optics leapfrogged in the twentieth century following several hundred years of scientific discoveries since Isaac Newton's work on optics in the mid-seventeenth century. The subject of "lighting" is duly distinguished within the general field of optics. Lighting is specific to human vision that is entirely dependent on illumination of objects by means of natural and artificial light and thus only deals with light in the visible spectrum. **Optics** is the branch of physics involving the behavior and properties of light in general, within the entire optical frequency spectrum encompassing visible, ultraviolet, and infrared light.

As the optical sciences have undergone rigorous development and become relevant to many disciplines including astronomy, medicine, photography, and many engineering fields, such as fiber optics and optical communications, the scientific knowledge and quantitative characterization methods of lighting improved as a consequence. Nevertheless, lighting remains primarily practical and experiential; we utilize it every day and many of us have become naturally accustomed to having artificial light with certain illumination quality.

1.2.2 Introduction to Lighting Fundamentals

Daylight reaching us from the sun is the predominant form of natural light from which our familiarity with light and vision began. Vision has descriptions only in the presence of light, some of which include the color, size, and shape of objects we see and how bright they appear. Without light, we do not see our surroundings or any objects present in areas surrounding us unless the objects are some form of light source themselves, such as fire. Lighting fundamentals have been established to describe what we see and how well we see things in the presence of light. Understanding the lighting fundamentals is crucial for all lamp designers and technologists, in particular for those in the LED industry who are generally missing an illumination background. Because artificial lighting has been around for a long time, people already have a great deal of expectations for lighting quality, user friendliness due to ubiquitous standards, extensive product availability, and fairly low up-front costs.

We require vision and hence lighting simply to view our surroundings, perform visual tasks, and enjoy entertainment shows. Using quantitative parameters, lighting fundamentals offer descriptions of how well light sources illuminate for the purpose of our vision in the presence of those light sources. It should be noted that for practical purposes, such described human vision is not absolute but rather derived from what the average person sees based on some reasonable comparative statistics.

1.2.3 Quantitative Parameters of Lighting

1.2.3.1 Color Metrics

For most viewing applications, we use primarily "white" light, which Newton demonstrated to be composed of different colors with his famous prism experiment, basics of which can be easily repeated as shown in Figure 1.1. White light has a very broad spectrum of colors (i.e., it contains a broad range of frequencies

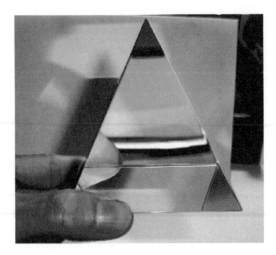

Figure 1.1. (See color insert.) White light from a compact fluorescent lamp is reflected from a shiny object and the light rays are captured by a camera after passing through a prism. The rays passing through the prism in this picture show the range of individual colors within the broad white spectrum of the fluorescent lamp.

or wavelengths). Since these optical frequencies vibrate at very high rates, our eyes or any other currently known detector cannot sense the fast-varying amplitudes of the light wave. Instead, our eyes assess an average optical power flow, which is referred to as **luminous flux.** Luminous flux is a scalar light power quantity measured in units called **lumen (lm).**

Illumination with broad-spectrum white light allows objects of various colors to be seen close to their inherent colors. This property of light is known as **color rendering.** White light, such as sunlight, contains all visible optical wavelengths or colors and we define sunlight as "pure" white light, as it is able to render all colors perfectly within our visual spectrum based on our definitions. Color rendering is quantified in a relative manner and is specified as a color rendering index (CRI). Although CRI is measured on a scale from 0 to 100, where 100 is considered to be ideal, low values are not meaningful in practice because they would not correspond to white light sources.

Sunlight has different intensity and hue, or tint, during the day from dawn till dusk and daylight varies accordingly, further changing with different atmospheric conditions that affect sky light scattering. This varying tint can be described with a parameter known as **color temperature.** The color temperature of a visible light source is the temperature correlated to that of an ideal black-body radiator whose emission radiation best matches the tint of that light source. (A black body is an idealized physical matter that absorbs all incident electromagnetic radiation.) It is thus more accurately known as **correlated color temperature (CCT)** and has the unit of absolute temperature measured in **kelvin (K),** which is occasionally also denoted as °K.

The CCT of a light source is quantified as the *surface* temperature of an ideal black body that emits light having the same color tint as the light source. An

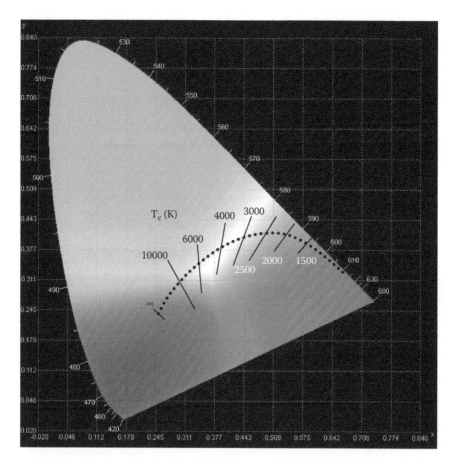

Figure 1.2. (See color insert.) An indicative diagram of the CIE 1931 (*x,y*) chromaticity space generated from a spectrometer by GL Optic GmbH using their GL SpectroSoft software. The dotted line here illustrates the black-body locus, which shows various points of constant CCT represented by the intersecting lines through it.

incandescent light bulb serves as an approximation of an ideal black body and its color temperature is essentially the same as that of its glowing filament—typically falling in the range between 2700 and 3000 K. The higher the color temperature is, the more bluish the tint turns, as seen in Figure 1.2, which shows the Planckian or black-body locus line in the indicative CIE 1931 (*x,y*) chromaticity space diagram; the black-body locus shown with a dotted line is intersected by several constant CCT lines. Such CCT values are only valid for sources whose (x,y) coordinates fall within some band defined by constant CCT lines.

The CCT of sunlight changes over the course of the day. Daylight that has a CCT of 6500 K has become the standard for various viewing applications. It is referred to as the D65 viewing standard. Daylight with CCT of 5500 K is the standard for photographic film.

Table 1.1. Color Temperature Chart for Familiar Lamps

Source	Correlated Color Temperature (CCT)
Incandescent	2700 K
Halogen	3000 K
Fluorescent—warm	3000 K
Fluorescent—neutral	3500 K
Fluorescent—cool	4100 K
Fluorescent—natural	4500 K
Fluorescent—daylight	6500 K

Note: The values in these tables are taken from the averages of several published sources including product handbooks from major lighting companies; thus, these CCT values are only nominal and considerable variations may be present among similar products.

Table 1.2. Color Temperature Chart for Natural Light, Common Light Sources, and Illuminated Screens

Daylight Type	Correlated Color Temperature (CCT)
Candle flame, sunset/sunrise	1850 K
Halogen	3000 K
Moonlight, xenon arc lamp	4100–4150 K
Horizon daylight	5500–6000 K
Vertical daylight	5100 K
Daylight, overcast	6500 K
LCD or CRT (cathode ray tube) screens	6500–9300 K

Note: The values in these tables are taken from the averages of several published sources including product handbooks from major lighting companies; thus, these CCT values are only nominal and considerable variations may be present among similar products.

During the night hours, we have become accustomed to warm light with yellowish hue—like that from candles and incandescent lighting; during the day, light with bluish hue, or cool light, is more effective—such as those from fluorescent and natural lighting—because it provides higher contrast among many colors.* Table 1.1 shows the CCT of various familiar lamps; Table 1.2 shows the CCT of various daylight, natural light, and a few other common light sources and lit electronic screens for reference. Because white LED lamps can be constructed with multiple narrow-wavelength sources and various color phosphors, the CCT can be varied according to what is desired.

Since the color rendering index is measured comparatively, the CRI of a light source is only meaningful when the CCT of the light source matches that of the reference. Thus, the CRI of an incandescent lamp is 100 by definition because

---

* A large number of applications support these inclinations since warm light (1850 K–3350 K) usage includes home, restaurant, studio photography and others that entail a relaxed atmosphere. Similarly, many task-type applications use natural lighting (5500K and higher); in particular, LCD and CRT screens offer the best visual acuity and color contrasts using a high-CCT white screen base.

it essentially *is* the ideal black-body source, which is the reference. Similarly, all natural daylight has a CRI of 100. In plain description, the CRI is the light source's ability to make colors look natural. The CRI, CCT, and chromaticity described by Figure 1.2, for example, form the basis of color description and metrics in lighting.

1.2.3.2 Luminance, Illuminance, and Spatial Light Distribution

Aside from color metrics, lighting fundamentals also include other important parameters that describe the illumination from light sources. Using the three following primary measurable parameters, one can explain how bright light sources are and how bright and pleasing objects will appear before our eyes, from various positions, when illuminated with certain light sources:

1. Luminance*

2. Illuminance

3. Spatial light distribution

The third parameter affects the distribution of light at the object surface resulting from the amount of incident light on the object coming from the various directions off the light source and other reflected surfaces. All of these parameters include luminous flux, which is quantified through direct detection of light power, weighted by a visual sensitivity function that models the average human brightness sensitivity. This **visual sensitivity function,** $V(\lambda)$, is wavelength dependent, as denoted here, with "λ"; it is also dependent on the amount of ambient light. Typically three such sensitivity functions are defined based on daylight, twilight, and darkness adaptations known as photopic, mesopic, and optopic luminous functions respectively [9]. The standard characterizations of light sources use the photopic weighted sensitivity function.

The first two parameters are related, as we shall see next. **Luminance** is the measurable quantity relating to actual optical power density produced by the light source in a certain direction. While it most often translates to visual brightness, it is different from brightness in the sense that brightness is only a physiological sensation and does not always have specific relations to optical power density. Note that this difference is not due to the human eye sensitivity functional variations described earlier; rather, brightness as a vision sensation may be affected by illusionary effects and surrounding light or shadows. Luminance and brightness may be used interchangeably for most cases, bearing in mind, however, that luminance represents the measured quantity not affected by optical illusions or adjacent light and shadows. Luminance is defined by optical power density or luminous flux density, per unit solid angle. It is used to characterize brightness of automotive headlights, projection lights, and various computer monitor displays;

---

* Although some may consider candela to be a fundamental photometric unit, it is not measured, but rather calculated from the measured parameter, luminance.

when used exclusively, luminance is usually not a suitable parameter for illumination applications.

Illuminance relates to the amount of optical power incident on a planar surface. It is typically used to characterize luminaires that provide illumination for objects and surroundings. The simplest way to determine whether we have adequate illumination for effective viewing is to measure the amount of light incident on a task plane or object plane, provided we are within some appropriate viewing distance from such a plane. This measurable quantity is known as illuminance, which is defined by luminous flux per unit area.

As mentioned previously, luminance and illuminance are related. This relation between these customary measured parameters is best can be illustrated in the simplest terms considering only normal incidence, as the following [10]:

$$E_v = \frac{L \times S}{D^2} \tag{1.1}$$

where E_v is the illuminance, L is the luminance of the luminaire, S is the surface area of the luminaire, and D is the normal distance from the center of the luminaire to the center of the illuminated surface. Equation (1.1) follows from the well-known inverse square relation that Swiss–German physicist Johann Lambert (1728–1777) helped formulate in the eighteenth century. The reader should note that the preceding equation assumes L is uniform, D is much larger than S, and ignores the cosine dependence of illuminance as well as the full integration over an object surface in order to illustrate here the simplest correlation among the two parameters.

The third parameter (i.e., spatial light distribution or **flux distribution**) describes how luminous flux from a light source is distributed over space, or often on an arbitrary surface. The surface may be planar or three dimensional. In space, this may be defined as a function, $\Phi(x,y,z)$, in Cartesian coordinates, or as a function in some angular coordinates. In practice, such functional values are measured on a point-by-point basis over the space of interest to create a map. The spatial distribution may also be calculated using ray optics approximation or other methods. The knowledge of spatial light distribution of light sources is important in lighting design to optimize visual performance and comfort within the entire field of view from various locations of interest, particularly when three-dimensional illumination of objects is of interest. It also allows designers to avoid excessively bright or dim areas and balance brightness. This parameter is also important for creating a uniform illuminance on a task plane or on displays, where it will have planar variations.

The lighting parameters discussed thus far are the basic parameters that characterize a light source or a set of light sources that can be used to provide one's desired illumination. However, there is a set of other related parameters that describe how well an environment is illuminated, what level of eye comfort is provided for the viewers, and how energy efficient the overall lighting system is. The measurements of energy-efficiency parameters described by lamp efficacy and luminaire efficiency will be discussed later in the chapter.

1.2.3.3   Summary of Lighting Metrics Used by Engineers and Manufacturers

Many lighting parameters are related and interactive; different combinations of them produce different results in various surroundings for human perception. However, using a primary set of lighting parameters as a basis for designing and manufacturing light sources and systems for illumination works best (summarized in Table 1.3). LED lighting engineers and manufacturers are particularly encouraged to incorporate *most or all* of these parameters simultaneously for developing LED lamps for various applications.

Lighting is still a developing field—in terms of both illumination engineering and the influence of human emotions on visual perception. It is also largely

Table 1.3. Basic Lighting Metrics for Lamp Designers and Manufacturers

Quantity (Symbol)	SI Unit	Unit Symbol	Dimension (Quantity Description)	Notes
Color rendering index (CRI)	None	None	1	Range: 0–100; 100 is best
Correlated color temperature (CCT)	kelvin	K	Degree K	Describes the radiation tint; various tints can be seen in Figure 1.2
Luminous flux Φ	lumen	lm	Visible light power unit with a correlation to electrical watt (F)[a]	Light power weighted by a human eye sensitivity factor
Luminance (L)	lumen per steradian per square meter	lm/(sr-m^2)	F/(sr-L^2)[b]	Units are also called nits; lumen per steradian is also known as "candle (cd)"
Illuminance (E_v)	lumen per square meter (lux)	lm/m^2 (lx)	F/(L^2)	Used for light incident on a planar surface
Flux distribution $\Phi(x,y,z)$	lumen	lm	Flux at a physical point or location	A map of lumen values over space of interest
Lamp efficacy	lumen per watt	lm/W	F/W	Used for lamp driven by electrical energy
Luminaire efficiency	None	None	1 (described in percentage or fraction)	Ratio of light power from a complete lamp unit including optically altering components to the bare lamp source light power

[a]   "F" denotes the flux dimension, not the unit. Visible light power (flux) or photometric power correlates to radiometric power by means of $V(\lambda)$; radiometric power is measured in watts—the same unit used for electrical power.

[b]   "L" denotes the length dimension, not the unit. A steradian is a unit solid angle denoted by "sr". The solid angle is defined as the two-dimensional angle in three-dimensional space that an object subtends at a point.

application dependent and thus, on occasions, additional parameters and effects as well as more complicated interdependency among parameters need to be considered than those presented here. In Chapter 4, we provide some increased details of these considerations; for additional information, the reader is also encouraged to delve into further reading on lighting design and specification parameters from the 10th edition of the *IESNA Lighting Handbook* [11].

1.3   Lighting Technologies

1.3.1   Introduction

Since the popularity of incandescent lamps marched forward in the early twentieth century, virtually all artificial lighting still continues to be driven by some sort of electrical method. Following the incandescent lamps, we have had various types of electric discharge lamps, starting in the 1940s. Among them are a variety of fluorescent and high-intensity discharge (HID) lamps, which all operate on the basic principle of light generation by sending an electrical discharge through an ionized gas.

1.3.2   Fluorescent Lamps

Fluorescent lamps are the most popular gas-discharge lamps that operate at a small fraction of atmospheric pressure. Fluorescent irradiance is produced by stimulating mercury atoms with an electrical discharge in a semivacuum enclosure, coated with phosphor, that give off ultraviolet light; the ultraviolet light is then absorbed by the phosphor that re-emits visible light in discrete blue, green, and yellow-orange spectral regions. Fluorescent lamps are much more energy efficient than incandescent lamps; however, they lack a broad spectral distribution and hence their CRI tends to be quite poor, particularly in the warm color zones. The fluorescent lighting technologies improved over many decades now produce a variety of more useful lamp shapes and sizes with longer life, faster turn-on times, more reliable internal discharge, and better CRI utilizing different phosphor combinations. The earlier family of these lamps came in larger tubes with linear, circular, and U-shapes as shown in Figure 1.3. These are widely used in commercial applications where bright illumination of large spaces is required, often with natural daylight-like color temperatures, to create a lively atmosphere. Certain lamps of these types are also suitable for residential uses such as in garages, basements, and workshops.

1.3.2.1   Compact Fluorescent Lamps

Compact fluorescent lamps (CFLs) have become a real contender to replace household incandescent lamps for a majority of the people around the world. With a great deal of technology improvements, they now combine the economies of fluorescent lighting with the comfort and versatility of standard incandescent lamps. The comfort is achieved with a warm CCT near 2700 K and a CRI of over 80, while cost savings are credited to an average rated lifetime of 6000 to 15,000 hours (based on 3–4 hours of daily usage on average, 7 days per week) and energy savings of up to 75%. CFLs screw into existing incandescent sockets

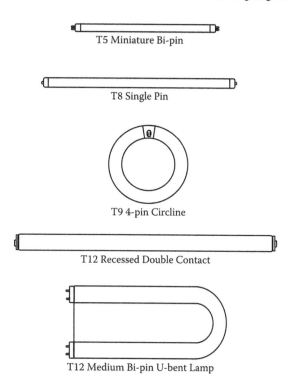

T5 Miniature Bi-pin

T8 Single Pin

T9 4-pin Circline

T12 Recessed Double Contact

T12 Medium Bi-pin U-bent Lamp

Figure 1.3. Fluorescent lamp shapes available in stores for standard uses. "T" stands for tubular, which is followed by the number that gives the tube diameter in units of 1/8 inch.

and come in a variety of shapes to illuminate homes, offices, hotels, restaurants, clinics, retail businesses, schools, and other places effectively. A small family of such CFLs found in stores is shown in Figure 1.4.

1.3.2.2 Toxicity

All fluorescent lamps contain some mercury. The Environmental Protection Agency (EPA) of the US government developed a test method known as the "toxicity characteristic leaching procedure" (TCLP) that classifies waste as either hazardous or nonhazardous for disposal purposes [12]. Fluorescent lamp manufacturers are directed by the US EPA to control their mercury content to pass such tests and recycle all such lamps regardless of their pass or fail status.

1.3.2.3 High-Intensity Discharge Lamps

High-intensity discharge (HID) lamps are used when higher levels of light over larger areas than those discussed for fluorescent light applications are required. Mercury vapor, metal halide, sodium vapor lamps, and xenon arc lamps comprise the HID lamp family. These lamps produce light by passing an electrical current through a gas or vapor at high pressure, which then produces a high-intensity discharge or arc of light. The process is very efficient and produces longer life lamps

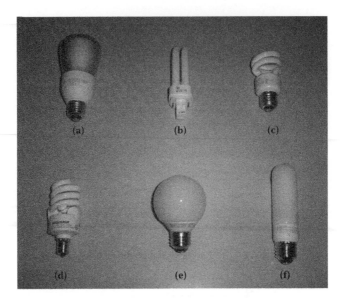

Figure 1.4. A small ensemble of compact fluorescent lamps available in regular stores: (a) 11 W R20 dimmable bulb in soft white; (b) 13 W two-pin tube in warm white; (c) 9 W deco twister in warm white; (d) 13 W minideco twister in 2700 K; (e) 9 W globe in 2700 K; and (f) 7 W long bulb in daylight 6500 K.

compared to fluorescent lamps. Their lifetimes are rated up to 24,000 hours. Typical CRIs for mercury vapor, metal halide, and high-pressure sodium are 20, 65, and 20 respectively.

Mercury vapor HID lamps have a typical CCT around 4500 K; the poor CRI of 20 confines its applications to outdoors only. Landscape lighting, dusk-to-dawn fixtures, roadway, parking lots, floodlight, and security use these lamps, all operating at a respectable efficacy of around 50 lm/W.

Metal halide HID lamps are used both indoors and outdoors because of their good color rendering, crisp white light near 4000 K CCT, and high efficacy of around 100 lm/W. They are extensively used in shopping malls, certain commercial buildings, film projectors, parking lots, airports, roadways, sports arenas, and for building flood lighting. Their electrical power ratings may range from 10 to 18,000 W.

High-pressure sodium (HPS) HID lamps are the most efficient HID lamps. When color rendering is not critical, HPS lamps are an excellent choice because lamp efficacy is over 120 lm/W, making them the most economical outdoor lamps of the current times. They are also more effective for keeping insects away than other lamps.

1.3.3 Incandescent Lamps

Many people still consider incandescent lamps to be most desirable to create personal-type atmospheres in such places as residences, restaurants, and hotel lobbies despite their very low energy efficiency and short lifetime ratings. Standard

Figure 1.5. A few incandescent lamps that are still favored by many users in the United States for various applications: (a) 50 W reflector in warm white; (b) 75 W omnidirectional bulb in soft white; and (c) 40 W interior lamp in 3000 K.

lamp life is generally rated between 750 to 1250 hours, depending on the filament thickness; thicker filaments last longer, but at the cost of lower light output because less heating or incandescence is produced from the fixed input current. Unlike fluorescent lamps, they can scale to very small sizes such as those found in Christmas tree light strands. Figure 1.5 shows a few variations of incandescent lamps commonly used as of today in the United States.

1.3.3.1 Halogen Lamps

Although halogen lamps are incandescent lamps, they are superior to their standard counterparts because they last longer while producing brighter and whiter light. This is accomplished in a halogen lamp by placing the tungsten filament inside a glass capsule filled with halogen gas, which allows the evaporated tungsten deposits to transport back to the filament instead of building up on the glass wall. Thus, the bulb is replenished and maintains cleaner and brighter light over longer times. The deposition process, however, repeats unevenly and eventually some parts of the tungsten filament turn weak and fail, ending the lamp's life. Halogen lamps are available in a variety of shapes, sizes, and power ratings and are used for general, task, outdoor, and automotive headlamp lighting applications. Figure 1.6 shows a few halogen lamps utilized for some of these applications.

1.3.3.2 Other Lighting Technologies

Other luminescence processes using various electrical methods are being considered for manufacturing lamps to compete with existing lamps. **Electron stimulated luminescence (ESL)** is one such technology that generates light by exciting phosphors with accelerated electrons [13]. Further demonstrations are still needed to assess its competitiveness against the current lamps.

1.3.4 Light-Emitting Diode Lamps

Light-emitting diode (LED) lamps operate based on electroluminescence in semiconductor materials. LED technologies have undergone many years of industrial and academic developments in various fields including display and communications. LED lamps are now available in limited quantities for general illumination,

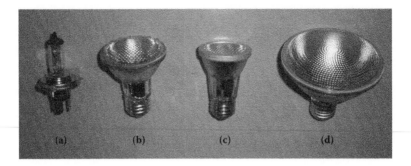

Figure 1.6. Variations of retail halogen lamps in the US market: (a) a standard automotive headlamp using 5 A/12 V or 60 W power input; (b) a 50 W and 525 lm rated PAR20 lamp; (c) a 60 W and 600 lm PAR16 lamp; and (d) a 60 W and 1100 lm PAR30 lamp. The b, c, and d lamps here are equivalent to 60, 70, and 75 W regular incandescent lamps, respectively, with regard to the total lumen output generated at the maximum rated input power.

primarily targeting replacement of 40 and 60 W incandescent lamps [14]. Because lamp prices are still steep and the technology continues to develop, with both cost and quality improvements changing in favor of consumers, end users have been reluctant to choose them for long-term use. Since LED lighting is the subject of this book, we will explore details about this technology throughout all the chapters. Figure 1.7 shows several retail LED lamps with the standard Edison base that is used in most fixtures.

The lighting technologies using electrical methods discussed here all have certain advantages and disadvantages in this present era. While there is a general inherent trade-off between lamp efficacy and CRI, there are additional trade-offs among several other parameters, some of which may be specific to applications. As the global need for lighting is readily increasing and lighting usage is becoming more versatile, it is important to understand these trade-offs and to utilize the most meaningful lighting solutions for each use in terms of economy, safety, and sustainability in both the near and long terms.

1.4   Understanding Illumination

Illumination from daylight and other natural light has trained our eyes to view things in particular ways. We recognize shadows, three-dimensional objects in nature, movements, and color in ways that have become our common vision. Our vision is also fundamentally shaped by all natural light including day, twilight, and night lighting with respect to motion sensitivities and fields of view. For example, in our retina, the rod and cone receptors that are responsible for peripheral and direct view, respectively, are variably sensitive to light levels, color, and motion detection during day, twilight, and nighttimes. Cones are situated densely at the center of the field of view and are most sensitive at high levels of light; in contrast, rods are responsible for our vision at low light levels and are most densely situated

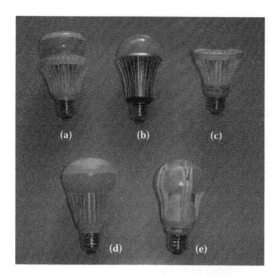

Figure 1.7. An assortment of LED lamps that penetrated the retail market in recent times: (a) a dimmable, 850 lm, 3000 K lamp generates near-ambient light using 13.5 W, (b) a 450 lm, 3000 K lamp generates downlight using 7.5 W; (c) an 8 W, 350 lm rated dimmable PAR20 lamp; (d) an 8 W, 450 lm rated lamp in 2700 K; and (e) a 2700 K ambient bulb rated at 12 W and 820 lm. The a, b, c, d, and e lamps shown here are equivalent to 60, 40, 40, 40, and 60 W incandescent lamps, respectively, with regard to the total lumen output generated at the maximum rated input power.

at the retinal periphery. The rods are also more sensitive to motion, which helps us to be more alert in the dark if a predator moves in the periphery of our vision [15]. Our eyes respond to daylight differently, as mentioned earlier in this chapter, based on photopic vision with respect to the color spectrum. When light levels are high, typically more than 3 **nits** (units of luminance), our vision is termed "photopic" regardless of whether the ambient light is natural or artificial [16].

As people have developed various types of artificial lighting over many years, our vision has also been influenced and has become accustomed to some standards based on illumination provided by such lighting. Understanding illumination is about understanding such bases or standards of an average person's vision that involve many lighting parameters, some of which can be currently described more scientifically than others. Interestingly, the field of illumination involves science and human emotions, as well as human subjectivity.

1.4.1 Lighting Parameters: A Deeper Look

In addition to the fundamental lighting parameters discussed in Section 1.2, a range of other parameters help provide a more complete description of illumination created by light. These include color balance, acuity, contrast, glare, interference or disturbance, time variance and adjustments, aberration, and polarization effects. When adjusted properly, these enhance viewing quality, provide eye

Box 1.1   Lighting Application Categories

Because of the way modern living has evolved, it is helpful to categorize light sources for two main applications: (1) illumination and (2) viewing. **Illumination application** is defined when we view things in secondary light that is reflected off the objects that we view. **Viewing application** is defined when we look directly at the illuminated surface, such as an electric sign or a computer screen. Figures 1.8 and 1.9, respectively, show examples of illumination and viewing applications that both use LED lamps.

A **luminaire** is a complete pluggable light source unit that is used to provide illumination of general objects; displays and signs are illuminated objects that we view directly and they are not meant to provide illumination. In many cases, signs and displays are illuminated with light sources that are luminaires and therefore this category is not unique to the first. Nevertheless, the requirements and specifications are different for these two categories of lighting products.

Figure 1.8. Several objects illuminated in a kitchen using an under-cabinet LED-arrayed lamped system.

For illuminated signs and displays, specifying luminance is most important and customary to determine whether or not they are suitable for viewing. Since we view them directly, it is necessary and sufficient to know the display brightness along the viewing direction to determine its effectiveness. (For some displays, the color requirements and specifications can be quite complicated and are not discussed here.) Because most displays tend to be two-dimensional viewing surfaces, the full spatial light distribution of luminaires that illuminate them is generally not relevant, although uniformity within the viewing plane is important.

It should be noted that there are two kinds of illuminated signs and displays: static signs that are internally or externally illuminated with lamps, and dynamic signs and displays (also known as digital signs) in which the viewing screen itself may be emissive and the content on the screen can be changed electronically. The topics relevant to dynamic digital signs and various indoor and outdoor electronic displays are very broad and intensive; some include sophisticated color creation and management for full motion, as well as color displays utilizing many different software and hardware technologies [17–19].

Figure 1.9. An RGB LED-lamped billboard alongside New Jersey's I-287 highway. It is built by Daktronics and operated by ClearChannel, which is responsible for rotating eight advertising messages at 8-second intervals.

comfort, and reduce energy consumption. A good lighting system design should take into account as many of these parameters that are deemed relevant for specific illumination applications. It is important to note that such design may get rather complicated because many of these parameters have complex interdependencies. As such, appropriate design methods should be adopted based on certain desirable illumination to be created, rather than simply designing for the amount of light a system would provide.

Light sources and luminaires have been developed by engineers and scientists and used by lighting designers for illumination applications for over a century. As new lighting technologies and applications are developed, understanding general illumination as well as novel uses is necessary for both incumbent lighting professionals and newcomers.

Providing the right illumination means shining the "right" light on an environment. When done successfully, it creates an effective and enjoyable viewing experience. The "right" light is usually best specified via the parameters in Table 1.3 and sometimes with other augmented parameters; it ensures the appropriate color balance, acuity, and contrast while minimizing or, when possible, eliminating glare, interference, and excessive brightness. The color of light is a very important aspect of illumination. The color temperature of lamps describes how the atmosphere appears; it helps provide an ambiance or mood created by light. Consistent color temperature should be maintained throughout a space utilized for themes such as within an entire ballroom, or a kitchen, or even a warehouse.

With respect to light level, or optical power quantity, it is flux distribution, $\Phi(x,y,z)$, rather than total luminous flux, that is the more important parameter that specifies illumination to a significant degree. Flux distribution depends on the shape and size of the actual light source elements or their combinations within a system. It provides the optical power levels in specific locations and the various light levels in these locations create the illumination in particular ways. It is best to use flat sources to illuminate flat surfaces, particularly when the surface to be illuminated is near the light sources; similarly, it is best to use round or round-equivalent sources to illuminate surfaces of all shapes because varying light levels are necessary on three-dimensional (3-D) object surfaces in order to view them naturally. Round or equivalently round light source elements produce nearly omnidirectional light for general illumination purposes. Unfortunately, it is difficult to produce round-shaped source elements for many lamp technologies, including incandescent and gas-discharge lamps. It is significantly more difficult to produce round or equivalently round-shaped LED light source elements.

However, deviating from flat source surfaces and utilizing some appropriate curvatures in the source shape still produces substantially uniform light in many directions. By incorporating many different periodic spatial curvatures in a single source, such as in the tungsten filament coil and in the CFL twist structure, omnidirectional light is effectively produced. The sizes of these source elements are also large enough to generate several hundreds of lumens to illuminate personal spaces such as rooms in typical houses, many restaurants, and offices where ambient lighting is desired.

1.4.2   Parameters Important for LED Lighting

Current LED technologies produce very thin, small, and flat light sources. Such shapes produce very directional light from small areas—usually a flat surface of only 1 mm × 1 mm. They are well suited to illuminate flat surfaces immediately next to them or very nearby, such as backlighting LCD (liquid crystal display) screens and lighting small retail merchandise displays placed on shelves. The directional light output from current LED sources is also satisfactorily utilized in flashlights, spot lights, and various accent lights. Because LEDs produce light based on electroluminescence properties of semiconductors, the color of light is determined by the material's finite and small energy-gap properties and consequently the emitted light consists of narrow wavelength bands bearing properties of associated single or near-monochromatic colors. In order to make LED lamps amenable to general illumination, innovations involving complex technologies will be required to achieve larger size lamps with near omnidirectional radiation and higher color quality features. These topics will be discussed in more detail in other chapters of this book.

1.4.3   Units of Measurement

The discussion thus far hopefully has demonstrated that understanding illumination requires the knowledge of spatial, spectral, and temporal parameters of light produced by various sources and how human beings perceive them. Quantifying many such parameters requires different units of measure, which form the basis of photometric characterization. Many people have often wondered why so many different units are necessary in lighting science; others are frustrated by the inability to convert between units that are each unique. For example, although lumens and candelas both are optical power quantities, they do not have the same units and one cannot be converted to the other. The parameter "lumen" does not specify where the unit amount of optical power is located; however, the parameter candela specifies that the unit optical power amount (i.e., 1 lm) is within one unit of solid angle or 1 steradian.

Often, in layman's terms, light or other physical quantities are described in manners that are ambiguous or incomplete. For example, the adjective "heavy" may be used to describe weight or density, but the two are fundamentally different. Likewise, people often use the word "bright" to describe a light source that produces a high luminous flux; they also term an LED or a laser "bright" because they concentrate high luminous flux into very small spots compared to other light sources. Although, in both cases, the power outputs are measured in lumens, these light sources need to be described using additional parameters involving different units. Usually different applications are supported with distinct cases. For example, one can "brightly" illuminate typical offices using an array of many recessed linear fluorescent lights to aggregate a high luminous flux. On the other hand, a laser that is blindingly bright in one direction (a laser beam has a very high luminance) cannot illuminate a room very well because it has very low luminous flux and its light is confined only to a very small region.

Illumination by light sources takes shape because of the ways light propagates through three-dimensional space, obeying properties such as divergence, focusing, reflection, refraction, transmission, absorption, and diffraction—all of which are wavelength dependent. These light properties affect illumination within a defined space when light gets subjected to such physical changes as spreading, combining, becoming concentrated, and reflection off polished or matte surfaces. Such properties and their codependency make the number of fundamental light measurement parameters or metrics rather large. Consequently, the numbers of quantities and units that represent them are also large. Understanding illumination in practical terms involves comprehending this large set of light parameters as well as being able to calculate and measure them to a great extent.

1.5   Understanding Energy Efficiency

1.5.1   "Green" Energy Solutions

Human society in many parts of the globe has become exceedingly productive since the advent of practical artificial light. Unfortunately, regions of the world without artificial light have not been able to keep up with modern productivity levels. While it is encouraging that artificial lighting is now reaching many developing nations, these places as well as the developed countries are consuming ever more energy for lighting and other modern conveniences that serve to improve people's lives. Thus, planning for reduction of energy consumption through more energy-efficient technologies and utilizing more renewable energy sources are two timely and important related topics for this book.

The lighting industry has been recognized as one of the most important fields for targeting energy-efficient solutions because lighting accounts for more than 20% of total global energy consumption. For many countries, it has even become a mandate to disallow the usage of certain incandescent lamps that consume higher energy in place of energy-efficient lighting. CFLs as well as LED lamps and luminaires are being pursued as suitable replacements, both of which promise to save 75% or more electrical energy consumption over the current usage of incandescent lamps for certain applications. This savings, however, does not translate to CFL and LED lamps being 75% more *efficient* than their incandescent counterparts!

In order to make the right choice for replacement lamps, it is crucial to understand some fundamentals of energy efficiency with respect to light generation and utilization. In the case of most practical lamps, electrical methods are widely used as discussed in Section 1.3. Therefore, in the context of lamps, it is important to understand how efficiently electrical energy is converted to light energy in end-to-end lighting systems of interest. Since lighting applications are very diverse, understanding efficiency involves terminologies, scaling, performance trade-offs, and economic issues. All of these factors need to be considered carefully in order to estimate the real efficiency of a lighting system over time.

Energy efficiency of electric lamps is described by lamp efficacy (also known as luminous efficacy) because, while light is produced using electrical energy, it is defined as energy evaluated by the human eye. Visible light correlates to the human eye's visual response and hence is not defined as energy in the same manner as other forms of radiation. Electrical energy is instantaneously driven into a lamp as electrical power measured in *watts,* whereas the lamp's light output power is quantified in *lumens,* whose calculation requires the knowledge of the spectral power distribution (SPD) of the lamp as well as the visual response of the eye. To determine lamp lumens, the light power at each wavelength in the visible spectrum is multiplied by the equivalent $V(\lambda)$ eye sensitivity value as discussed earlier; all such values are then summed over the full color spectrum to find the total lumen output. This is stated as

$$\text{Lamp lumens} = C \cdot \Sigma \text{ lamp power (in lumens) } (\lambda) \cdot V(\lambda) \cdot \Delta\lambda \qquad (1.2)$$

where C is a constant quantity to balance units.

Although $V(\lambda)$ depends on ambient light levels, lamps' lumen output is usually rated for some average light level condition in the photopic range. Determination of a lamp's lumen output allows one to quantify the lamp's efficacy by taking the result from Equation (1.2) and dividing it by the total electrical power used to drive the lamp. The energy efficiency characteristics of lamps may be described using both efficiency and efficacy parameters, but the two terms are *not* interchangeable.

1.5.2 Luminous Efficacy versus Luminous Efficiency

Luminous efficacy is a conventional measure of how well an electrical lighting element produces visible light. It is a measure of how much luminous flux is produced per unit of electrical input power and therefore is measured in "lumen per watt" (lm/W). It should not be confused with luminous efficiency, which is given in percentages where both power units must be either in lumens or watts, depending on what efficiency one is interested in describing. For most traditional lamps and luminaires, such as those comprising incandescent and fluorescent lamps, luminous efficacy provides the lamp efficacy and luminous efficiency provides the luminaire efficiency. **Luminaire efficiency** is of interest when we ask, for example, how much light is still available when a lamp is placed in a fixture where some light is being hindered by the fixture. Consequently, it is determined as the following, from the ratio of two light powers measured in lumens:

$$\text{luminaire efficiency } (\%) = \frac{\text{luminaire output (lm)}}{\text{lamp output (lm)}} \times 100 \qquad (1.3)$$

If both electrical and light power units could be expressed in watts, from a luminous efficiency parameter we would get a complete or absolute idea of what percentage of electrical energy we have been able to convert to optical energy at one instant of time. However, since visible light power is defined only using lumens and accounts for eye sensitivity, we need the efficacy quantity to provide a measure for energy efficiency. This efficacy is a figure of merit that tells us how many lumens are being generated per unit of electrical power, which then allows us to determine how many total lumens we can get out of certain electrical wattage going into the lighting system, provided a linear scaling is meaningful for the entire system.

These concepts are important for all lighting in general. For LED lighting, the industry focus has been largely on increasing efficacy. However, we need simultaneously to apply a basic understanding of end-to-end system efficiency to help determine theoretical and practical efficacy limits and make valid and meaningful energy efficiency comparisons among different lamps and technologies. An end-to-end system efficiency, for example, may be affected by such parameters as total effective lumens in desired areas, dimmability, thermal management schemes, and others.

1.5.3   Determining Maximum Efficacy

In order to determine the theoretical efficacy limit for any lighting unit, one must first ask how lumens relate to watts or what the watt equivalent of a lumen is. From earlier discussions, you may recall that the answer depends on the wavelength of light. Since white light consists of a broad set of wavelengths, the answer is not very straightforward. Let us see how one might nevertheless determine some approximations.

One lumen (lm) is the equivalent of 1.46 milliwatts (mW) of radiant electromagnetic (EM) power at a frequency of 540 terahertz, or 5.40×10^{14} Hz, which corresponds to the middle of the visible light spectrum at 555 nanometers (nm), where the human eye is *most* sensitive [20]. An EM field power level of 1.46 mW is rather small for most applications; for example, the radio-frequency (RF) output of a two-way toy radio is several times that amount. Since in the visible region, 1.46 mW = 1 lm at maximum, it follows that 1 W = 685 lm at maximum, which is valid only at 555 nm. One watt equals fewer lumens at other visible wavelengths.

Therefore, the highest efficacy theoretically attainable, corresponding to 100% efficiency, is 685 lm per watt. This would be the output obtained if all the input power were converted to green light at a 555 nm wavelength, where the sensitivity peaks for the human eye. The maximum theoretical efficacy of any light source producing white light with its entire output power distributed uniformly with respect to wavelength within the visible spectrum is only 200 lm/watt. Such an "ideal" white light source would have a perfect CRI of 100 by its own right. By concentrating the output wavelength of a light source near the 555 nm point, one can improve the efficacy beyond what is possible with an "ideal" white light (light consisting of all visible wavelengths with equal amount of "lumen" power in each wavelength). However, such an attempt to increase beyond the theoretical efficacy of 200 lm/W by concentrating more light output power near the green region would decrease CRI. Thus, there is an inherent trade-off between efficacy and CRI for white light sources.

1.5.4   Efficiency of Light Sources

Although lamp product descriptions generally include efficacy and lumen output ratings, it is helpful to estimate energy efficiencies of various lamps. So let us look at efficiency levels of some light sources.

A typical incandescent light with an efficacy of 13 lm/W is about 2% efficient based on a rough estimation using approximate lumen-to-watt conversions and $V(\lambda)$ responses; thus, nearly 98% of the input electrical power turns to heat. In contrast, a typical CFL with 70 lm/W efficacy is about 10% efficient, for which nearly 90% of input electrical power eventually converts to heat. Certain linear fluorescent lamps have efficacies as high as 100 lm/W, thus making them approximately 14% efficient.

In an incandescent lamp, typically 98% of input electrical power converts into radiant heat that is dissipated naturally and therefore designers do not have to worry about heat removal. In contrast, for an LED emitting at 555 nm with 70 lm/W, almost 90% converts into conductive heat, which must be removed from the LED chip by design for optimal performance.

1.5.5   LED Luminaire Efficacy

As discussed previously in Section 1.5.1, for conventional light sources, the lighting industry generally speaks of lamp efficacy and luminaire efficiency, because the lamp is detachable from the luminaire fixture. However in the case of LEDs, this is not always the case. When the LED light source is fully integrated in a luminaire configuration, the efficacy must be that of the entire LED luminaire system, and luminaire efficiency provided by Equation (1.3) no longer has any meaning. In such cases, only one parameter suffices, which is the LED luminaire efficacy.

The LED luminaire efficacy depends on several types of efficiency factors; the product of all such factors yields the total efficiency, η_T, from which one can estimate the LED luminaire efficacy. The total efficiency, η_T, is defined as

$$\eta_T = \eta_{int} \cdot \eta_{ext} \cdot \eta_{dr} \qquad (1.4)$$

where η_{int}, η_{ext}, and η_{dr} are LED internal quantum efficiency, light extraction efficiency, and the driver efficiency respectively. LED luminaire efficacy is linearly proportional to the total efficiency and therefore maximizing the efficacy requires maximizing all three factors in Equation (1.4). Further, if any of these factors diminish or have poor performance, the luminaire efficacy also suffers accordingly. These will be discussed in more detail in Chapter 2.

1.6   The LED Industry: Current and Future Prospects

1.6.1   Worldwide Growth

When Nick Holonyak, Jr., demonstrated the first visible LED, in 1962, he produced only a small fraction of a lumen, but since then the LED lighting market has become enormously successful. It holds the promise to grow well into a market size of tens of billions of dollars in only another few years as further advances are achieved in light extraction efficiency, magnitude, and quality. LED light sources have captured global attention across several industries, governments, and scientific communities, and in the public domain as we witness LEDs illuminate household TV screens and light up many entertainment zones, buildings, and constructs with electronic billboards, signs, and decorative lights. The excitement is so strong that a few years ago Samsung called its newest TV model "LED TV," even when those sets were only LED-backlit LCD televisions.

The industry is colossal as well as incredibly diverse, covering illumination not only for certain necessities, but also for various new entertainment uses. Creative groups are finding many delightful applications to light up everything from such small items as toys, clothing, and various household objects to very large building facades. Because of such market diversity around the world, it is difficult to determine the actual size of the LED market worldwide or even for such fast growing countries as China and South Korea. But it does appear certain that the LED lighting market, which continues to grow rapidly, will experience high growth for years to come.

1.6.2   High-Brightness LEDs

One way the industry separates the essential or traditional market from the entertainment market is through the size of the high-brightness, typically white LEDs used for LCD screen backlights, sign illumination, electronic message centers, refrigerator lights, display lights, outdoor lights, and automotive and various other niche lighting markets. This market segment, known as high-brightness LED (HB-LED), has now grown to over $10 billion in 2010, easily outstripping the laser market. Several manufacturers of HB-LEDs now commonly produce nearly 100 lm with 0.75 W from a single, packaged LED emitter for commercial uses, with some producing in their R&D laboratories even twice the optical power from single-color and white LEDs applying comparable electrical drive powers.

1.6.3   LED Applications

Currently, LED lighting markets are primarily serving three types of niche applications: (1) displays or signs that are directly viewed; (2) illuminating smaller spaces and only nearby objects, such as inside refrigerators, for merchandise display, or task lighting; and (3) outdoor lighting for streets, garages, etc. They all bear some common features in that they all illuminate flat surfaces or small, proximity objects; further, white light color quality requirements for these applications are usually not very stringent.

For general lighting, however, illuminating large spaces and three-dimensional objects with good color quality is essential. Thus far, LED replacement lamps for such applications are still underperforming the fluorescent and incandescent lamps as evaluated by the US Department of Energy (DOE). The costs of these LED replacements are also prohibitively higher.

Experts believe LED lighting will be the choice for essentially all lighting sometime in the future, primarily because LEDs have already proven higher efficacy than their fluorescent and incandescent counterparts at the source level (albeit within a small scale). Further, their theoretical efficacy could still be improved by at least a factor of two—even at higher drive currents. Although this is true, current LEDs are flat, discrete modules and higher efficacy translates to higher luminance or brightness only for the small modules, which does not easily scale to a larger lamp source for omnidirectional illumination, without innovative secondary optics.

Scaling an LED lamp to provide, say, 1000 lm omnidirectionally with a high CRI of over 90, from the size of a household incandescent or a compact fluorescent lamp (CFL)—while providing an adequate thermal management scheme—is quite a difficult engineering feat. This is particularly so when the cost of such a lamp needs to be roughly the same as that of a CFL.

However, LEDs are clear winners for distant-viewing electronic message centers (EMCs) that use the RGB (red, green, and blue) modules to generate full-color video displays or still images digitally. LEDs are also the preferred choice over other lamps for certain sign illumination applications. Signs, displays, and EMCs are directly viewed and therefore need to be bright enough for us to see their content, but they need not illuminate anything else. Back- or edge-illuminated LED channel letters and cabinet signs use the lamps in proximity to the sign faces and hence adequate surface illumination from an array of small LED modules is feasible. LED lamps' directional nature prevents them from illuminating large spaces or areas farther away. But this directional nature makes them more suitable for retail display lighting (jewelry, for example), task lighting, accent lighting, and for some street or down lighting. LEDs are not inherently directional but such light distribution results from their flat geometry produced by the practical wafer manufacturing process. It would be difficult to manufacture LEDs in semiconductor materials any other way, although some improvements in chip design may make them more suitable for somewhat wider angle viewing.

Box 1.2 Organic LEDs

Thus far, only the LEDs produced using inorganic semiconductors, which currently have a much larger market than that of organic LEDs, have been discussed. LEDs made from organic materials such as polymers are known as "organic LEDs," or "OLEDs." OLEDs are used to illuminate various small displays such as those found in mobile phones. OLEDs have two categories of products: passive matrix (PMOLED) and active matrix (AMOLED). AMOLEDs have integrated circuits that can control the OLEDs in the same device layout. OLEDs, as a single light source unit, can also be manufactured to produce much larger emitting areas than LEDs can, which along with an AMOLED's active circuit control features, have the ability to be used as self-emissive screens rather than just to illuminate LCD screens from the back or edges. Such OLED TVs have similar high-contrast benefits of plasma, but are much thinner and produce beautiful, bright, and saturated colors. Although some OLED laptops and TVs are available in very low quantities in the market today, they are rather small and still suffer from thermal instability and lower efficiency. Scaling up OLED products has proved to be quite challenging.

One of LED lamps' great benefits is that they are compatible with digital technologies. Thus, unlike with traditional light sources, lamp designers can incorporate intelligence in the products to control consumption and appearance according to various needs. The lamps can also be used as communication devices. Despite certain challenges of LED technologies, these lamps can potentially offer much more than illumination for a wide variety of applications. Because of the vast usability, LED usage will continue to increase as scientists and manufacturers work toward overcoming their challenges.

1.6.4   Challenges and Limitations for the LED Industry

It is plausible for the highly accelerated growth of the past decade to continue for the next few years, because the majority of this market will continue to include LED-illuminated display screens for a variety of industrial and consumer applications. LEDs are natural for such uses because they are structurally flat, small, integration friendly, and electronically controllable. LED enthusiasts project that its market will soon proliferate broadly into general lighting, eventually replacing widely used lamps in residences and commercial buildings. Such projections are unlikely to materialize unless LED lamps can be designed suitably for large space illumination with good color quality and uniformity and can be manufactured cost effectively.

The very nature of the current LED technology that makes it successful for such uses as back-illumination for displays and task and accent lighting also makes it unsuitable for ambient lighting. At the core of this nature lies the fact that LEDs produce light within single or narrow wavelength bands from a small and flat surface, resulting in highly directional light distribution. Further, CRI and CCT color parameters tend to vary widely due to variations within the same technology platform as well as due to cross-utilization of significantly varying platforms, when large quantities of LEDs need to be assembled for certain high-lumen applications.

Combating these challenges requires understanding of illumination fundamentals and techniques for controlling and scaling lights emanating from LED chips to provide the same or better ambient illumination than what we have been accustomed to for over a century. Such understanding of principles and methods is crucial for LED engineers to build highly efficient replacement lamps for general lighting, which currently dominates the lighting sector's global energy consumption.

Although the LED industry has enjoyed great advancements in generating light at the chip level, it generally displays insufficient understanding of large space illumination and 3-D object illumination. This misunderstanding comes to light when one recognizes that many manufacturers of LED replacement lamps for tubular fluorescent lamps have simply taken existing tubular structures and placed discrete and flat LED lamps all around them. However, illumination from such replacement lamps is far less desirable than those we currently use in offices and many other commercial buildings. Similar ineffective approaches are often seen in the industry where the most common A-line lamp (i.e., the Edison bulb) is imitated with an LED replacement lamp that consists of a flat, common substrate containing several LED chips, while only the enclosure mimics the shape of an Edison bulb. Although the external structure appears similar in these LED replacement lamps, the illumination from these lamps is far different from their existing fluorescent and incandescent counterparts. This is because comparable illumination is only generated when the actual light sources or source elements are optically equivalent or when adequate external or secondary optics are used to obtain equivalent illumination.

In order to produce such optical equivalency using current LED structures, some sophisticated and novel optical designs must be adopted. These may include physical optics, such as diffractive optics or free-form optics, or integrated optics, such as guided-wave optics technologies, in order to control and scale light to provide desirable large-space illumination that many incumbent lamps now offer. Accurate simulation of such designs using appropriate numerical techniques is necessary to predict the light distribution behavior of such LED lamps. Finally, the simulated designs need to be verified through photometric measurements of the parameters listed in Table 1.3.

Certain challenges in the LED lighting industry are situational since LED science and technology sprang out of solid-state physics and semiconductor device engineering (both of which are intense fields for optoelectronics but not for illumination). However, while lighting scientists and designers understand and appreciate illumination better than traditional LED engineers and scientists, lighting experts tend to be unfamiliar with LED science and technologies. In order to bridge the gap, multiple academic and industrial disciplines need to merge to develop design, testing, and manufacturing platforms that can appropriately address the unique challenges of producing desirable LED light sources.

The current LED industry also suffers from a number of broad inconsistencies in terms of product quality, costs, and characterization standards. To some degree, these inconsistencies stem from the nature of emerging technologies that generally need to experience a maturing phase. But, specifically, the lack of uniformity is a result of wide variations in light and electrical characteristics of LED chips produced from the same wafer during manufacturing. In contrast, the variations in other existing electric light sources are much smaller, and this has set the expectations for the general lighting industry. Large variations in LED lamps result from a complex set of compound semiconductor material growth, processing, and fabrication issues that are difficult to control within strict ranges.

Such inadequate control variably affects each LED chip's optoelectronic and thermal properties because they are intrinsically interdependent, leading to a laborious binning process that winds up producing far fewer high-end LED chips than low-end ones. As a consequence, high-brightness and high-quality LED lamps are currently prohibitively expensive for broad usage. On the other hand, the vast number of lower end chips are not discarded, but rather sold to other vendors who package them into lamps and sell them at much lower prices for other markets.

The absence of standards for LED lighting products and characterization has been apparent for some time and is still one of the key concerns for luminaire and lighting designers as well as end consumers. However, important progress has been made in recent years by standards bodies formed by the American National Standards Institute (ANSI), Illuminating Engineers Society of North America (IES), National Electrical Manufacturers Association (NEMA), and Underwriters Laboratory (UL) and supported by the US DOE. These include *ANSI C78.377-2008* (chromaticity specifications), *IES LM-79-2008* (methods for electrical and photometric testing), *IEM LM-80-2008* (method for measuring lumen depreciation), and *UL 8750* (safety standards). Other guidelines, recommendations, and projections documents are also available for LED lighting from these organizations [21].

Although these standards and white papers offer significant technical support to the industry, the development of LED replacement lamps for general lighting will still require further improvements in platform technologies as well as novel and cost-effective solutions for ambient lighting. While advancing the compound semiconductor industry for lighting will improve supply chain platforms that can enable higher lamp efficacy and more product uniformity, understanding the necessary lighting fundamentals will still be needed to develop LED lamps that can match and outperform the incumbent lighting solutions. These lighting fundamentals and LED lamp design methods may be established by developing underlying concepts, rigorous simulation capabilities, and novel design techniques. The remainder of the book is dedicated to providing the knowledge necessary for understanding LED illumination; in order to construct practical and competitive LED lamps, we must learn about the behavior of light in scientific terms, utilize a set of rigorous design and simulation techniques, accurately evaluate their photometric properties, and become familiar with some new design concepts and methods.

2 LED Lighting Devices

2.1 Introduction

This chapter elaborates on the core or the engine of an LED lamp, which is a particular type of semiconductor diode that produces light when it is forward biased. In the 1950s, the development of inorganic semiconductors was rapidly expanding for electronics as scientists and engineers recognized the potential of p–n junction diode devices for compact integrated circuits. Around this time, many researchers also delved into theoretical and experimental investigations of optical properties of these junction devices in a variety of semiconductor materials, looking to demonstrate efficient radiation emission and explore interesting applications.

A series of discoveries involving optical emission from various compound semiconductor diodes followed within the next decade. The first of such declaration came in 1955 from Rubin Braunstein of the Radio Corporation of America (RCA), who reported on the observation of infrared emission from simple diodes constructed from gallium arsenide (GaAs), gallium antimonide (GaSb), indium phosphide (InP), and silicon-germanium (SiGe) alloys at room temperature and at 77 K [22]. The first *infrared* LED patent, however, was awarded in 1961 to Robert Biard and Gary Pittman from Texas Instruments for the demonstration of optical emission by applying electric current to a GaAs diode [23]. During this time, the goal of many semiconductor researchers became one that would extend infrared LED development to demonstrate a diode *laser*, which turned into a heated race in the summer of 1962 [24–26]. By the fall of 1962, groups at IBM, MIT Lincoln Laboratory, and GE succeeded in demonstrating laser diodes, setting the stage for fiber-optic communications, advanced medical technologies, and other noteworthy fields.

Among them stood out Nick Holonyak, Jr., who was persistent in working with semiconductor materials that would emit light in the visible spectrum, deviating from the others, whose interests garnered around materials only emitting in the infrared region because they were easier to produce by more conventional means. While other groups succeeded in demonstrating infrared laser diodes by the fall of 1962, Holonyak's own discovery turned out to be twofold, implications of which went beyond what others then might have imagined: the first *visible* laser diode and the first *visible* LED! For others, the "visible" characteristic did not play any role because demonstrating the first diode laser of any kind had been the sought-after grand prize; therefore, it made sense for them to utilize GaAs that was available off the shelf, thus avoiding the difficulty of using any alloy material they did not know how to produce. Despite facing much criticism from others, Holonyak chose gallium arsenide phosphide (GaAsP) because he was sure he could produce diodes from such alloy materials. When he managed to get his GaAsP diode to operate as a laser, it was lasing visibly in red in contrast to his competitors' GaAs diode lasers that produced infrared emission. Because GaAsP has a higher energy gap than that of GaAs, it emits light with higher energy, reducing the wavelength of the emitted light from infrared to red. Holonyak proceeded with another step: While experimenting with adding more phosphorus, he also showed that, although light emission efficiency had dropped to stop the laser's coherent emission, there was still enough incoherent visible red emission from his diode—and that was indeed the first visible LED!

Nick Holonyak and his co-workers continued their work to produce various stable III–V compound semiconductor alloys with the necessary energy gaps (also known as bandgaps) to emit light in other visible wavelengths. Holonyak is seen as the "father of the light-emitting diode" who apparently had the early vision for semiconductors to provide energy-efficient lighting for everyday usage. Becoming familiar with the history behind the LED discoveries and the ensuing development of optoelectronic technologies in inorganic semiconductors in the following decades perhaps enriches the understanding of the apparent success of LEDs in the lighting industry.

2.2 Basics in Semiconductor Optoelectronics

Nick Holonyak's approach of tuning phosphorus in GaAsP set off an important research area of bandgap tuning with alloy compositions to produce spectrally diverse optical emission from compound semiconductor materials. While this accomplishment drove the "red" (includes orange, amber, and yellow shades) LED and laser diode industries for three decades, it was not until 1993 when Shuji Nakamura, from Nichia Chemical Corporation in Japan, demonstrated the first high-brightness blue LEDs using AlInGaN alloys that opened the door for the development of white LED lamps [27]. This discovery was another major milestone for the semiconductor lighting industry because it was apparent then that light can be generated reliably and efficiently in the whole visible spectrum using compound semiconductor alloy devices.

For fiber optic communications, *infrared* lasers, modulators, and receivers, which utilize GaAs and InP compounds and their alloys, became primary interests [28].

Indeed, for visible LEDs (and laser diodes), two separate base material systems became dominant: (1) GaP-based alloys for red, orange, and yellow LEDs, and (2) GaN-based alloys for blue and green LEDs. These two compounds belong to the III–V groups, which have many alloy compositions that can range from AlN and InAs to GaAs and InP, producing wavelength in the visible, near infrared, and ultraviolet spectra. These alloy materials with different compositions yield different bandgap energies. GaP and GaN are suitable base materials for producing ternary and quaternary alloys such as InGaP/AlInGaP and InGaN/AlInGaN; these allow desired bandgap tuning as well as creating alternate quantum well (QW) and barrier stacks to increase internal quantum efficiencies in visible LED devices.

2.2.1 Light Emission in Semiconductors

Compound semiconductors such as those described here have direct bandgaps; that is, their peaks and valleys of valence and conduction energy bands are aligned vertically as shown schematically in Figure 2.1(a). Such alignment allows little or no energy to be wasted when electrons undergo transitions from one band to the other to create photons [29]. Figure 2.1(b) shows the simplified

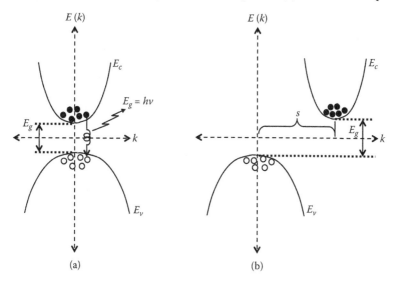

Figure 2.1. Simplified view of the band-edge structures of (a) a direct bandgap semiconductor where the photon energy equals the bandgap energy; (b) an indirect semiconductor where the photon energy equals the bandgap energy less the phonon energy, which is a function of the phonon wave vector, s.

band diagram of an indirect bandgap semiconductor, for which optical processes are less efficient because the electron undergoing a band–band transition shares its energy between a photon and a phonon. The phonon energy generates heat and further degrades the optical stability of the photonic device over time. Silicon (Si) is an indirect semiconductor that is very suitable for electronic devices; however, it has not widely proven to be a very efficient active material for optical devices.

This band-to-band transition is a part of the electroluminescence process through which an LED produces light when it is forward biased with an electrical input.

2.2.2   LED—The Semiconductor Diode

An LED is a semiconductor diode that is formed by joining p- and n-type materials together that have large charge carrier concentrations of holes and electrons, respectively. At the interface of the two material types, a junction is formed where electrons and holes exchange places through oppositely directive diffusion and drift motions initiated by thermal agitation. Such motions create a built-in potential (V_{bi}) energy whose strength is determined by the thermal equilibrium condition at which state the net carrier motions must come to a halt. This condition is satisfied only if the energy bands in the junction bend by the amount equal to V_{bi} so that equal probability (50/50 chance) of finding either type of carrier within the junction remains the same, which is required by the thermal equilibrium state. Mathematically, this is equivalent to requiring a constant Fermi level of energy throughout the diode material, as depicted in Figure 2.2(a), which shows the band bending for a p–n junction at thermal equilibrium with no external force applied to it. However, when a forward voltage bias, V_f, is applied to the diode, the band bending is reduced by V_f and the diffusion currents exceed the drift currents; now a net diffusion current starts to flow from the p-region to the n-region as shown in Figure 2.2(b). When a reverse bias V_r is applied, the band bending increases, as shown in Figure 2.2(c), and there is no diffusion current in this case—only a very small net drift current in the reverse direction, but none in the forward direction. Essentially, the reverse current flowing through the junction is negligible until the junction starts to break down at a very high reverse bias level.

Under forward bias, the larger electron and hole diffusion currents drive holes from the p-type material to the n-side, and drive electrons from the n-type material to the p-side in the junction region. This bidirectional injection of the minority carriers allows electrons and holes to recombine in the junction. Such an electron–hole recombination process can produce a photon when the electron releases energy by transitioning to the lower energy level. In direct bandgap semiconductors, the created photon has the same energy released by the electron, which is also the same as the bandgap energy. These quantities are related as the following:

$$E_e = E_g = E_p = h \cdot \nu \qquad (2.1)$$

where E_e and E_p are the electron and photon energies, h is the Planck constant, and ν is the photon frequency. Using the golden rule relation,

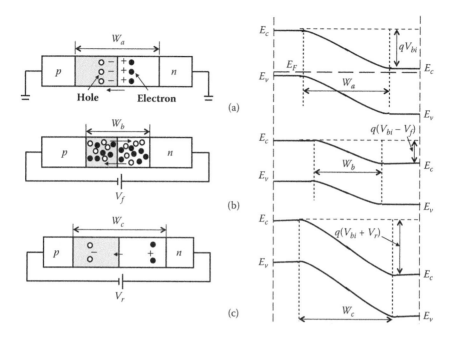

Figure 2.2. Schematic illustrations of a p–n junction (left side) and its corresponding energy band diagram (right side) for three cases: (a) at thermal equilibrium and without any external bias; (b) under forward bias, V_f, which reduces the band bending by V_f; and (c) under reverse bias, V_r, which increases the band bending by V_r. The junction width changes with applied bias, satisfying the condition $W_b < W_a < W_c$.

$$c = \lambda \cdot \nu \tag{2.2}$$

where c is the speed of light, we can write the wavelength of the photon, λ_p, as

$$\lambda_p = \frac{hc}{E_g}. \tag{2.3}$$

Thus, the photon wavelength (i.e., the color of light) is determined by the bandgap energy E_g.

2.2.3 LED Device Structure

In a simple LED device structure, the p–n junction, also known as the active region, is formed by epitaxially growing thin p-type and n-type alloy layers of the base material as discussed earlier in this chapter. Positive and negative polarity contacts are formed on these layers, as shown in Figure 2.3, through which the diode can be driven electrically to produce light.

In a common surface-emitting LED chip, the majority of the light escapes usually only from one surface because the other surface is blocked by its substrate or

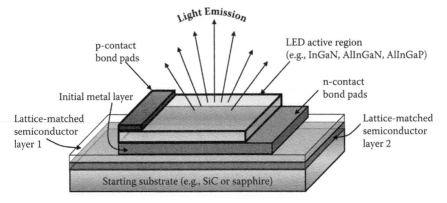

Figure 2.3. A schematic diagram of a surface-emitting LED chip or die showing its basic features and the direction of light emission. (The figure is not drawn to scale.)

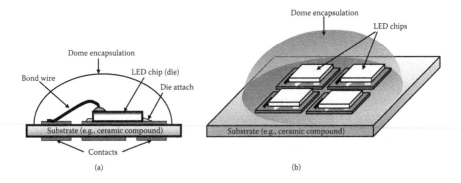

Figure 2.4. SMD module diagrams showing (a) the side view of an encapsulated SMD module with one LED chip inside; (b) a perspective view of an SMD module with four LED chips mounted on a common substrate.

some other added base on which a heat sink can be placed for removing excess heat through conduction. Such an LED chip or a cluster of them is then packaged into a single surface-mount device, known as an SMD, by placing them on a common thermally conducting substrate and enclosing them with a polymer-dome housing. Figure 2.4(a) is an illustration of an SMD module cross-section showing a single LED chip in the package, whereas Figure 2.4(b) is a three-dimensional (3-D) illustration of four LED chips packaged into a single SMD module.

The simple plastic dome-shaped enclosure in these modules usually does not alter the LED light output distribution. The light distribution from a single LED chip is approximately Lambertian, as depicted in Figure 2.5(a) and (b). It is possible to use this dome as a secondary optical lens element to shape the light distribution pattern to suit different applications.

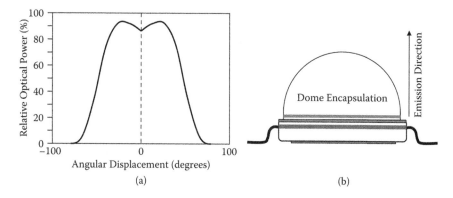

Figure 2.5. (a) A typical cross-sectional light distribution pattern of a single packaged LED is shown schematically; (b) the side view of the corresponding packaged module where the LED chip width nearly matches the maximum width of the light distribution profile shown in (a). The light distribution of such an LED module is approximately Lambertian.

2.2.4 White LED Configurations and Challenges

Most white LEDs in the market use blue LED chips made from GaN-based materials and some type of yellow or warm color phosphor coating, placed either directly on the chips or on the polymer or resin encapsulation, or even on the surface of the final enclosure. The blue light passing through the phosphor experiences the Stokes shift downconversion and appears as white light. This was first demonstrated by Shuji Nakamura and he was awarded the 2006 millennium technology prize for his blue and white LED inventions [30]. White LED lamps can also be constructed by combining light from several single colors, such as red, green, and blue (RGB) LEDs, by mixing appropriate color ratios via electronic tuning. RGB-type white LEDs are currently not desired over phosphor-coated GaN-based LEDs for most applications due to higher cost of multiple LED chips; their pronounced color inconsistencies, particularly due to aging; and added drive electronics.

With significantly higher levels of light flux and luminance compared to those from indicator lights, white high-brightness (HB)-LEDs have the highest demands in the LED lighting and display industries. Interestingly, the GaN-based devices, currently preferred for white HB-LEDs over other technology choices, suffer from significant reduction in internal quantum efficiencies at high drive currents, where the brightness levels of interest are produced. This phenomenon is termed "droop," which has been a controversial subject in the industry for many years as LED scientists and engineers have not agreed on any common cause for it. Various groups proclaimed different physical causes for droop with accompanying experimental findings [31–33]. The droop behavior essentially states that as the injection of current is increased, after a certain point there is an increasing amount of nonradiative electron–hole recombination or a decreasing amount of radiative recombination in relation to the drive current. The controversy is over what causes the net radiative recombination efficiency to drop at high currents.

Some have claimed that the Auger recombination process is the responsible nonradiative process. According to E. Fred Schubert and Dai et al., in nitride-based LEDs, the nonradiative processes become more dominant due to current leakage; they observed droop at injection current densities exceeding values that range between 0.1 and 10 A/cm² [31]. Other groups have also claimed that the droop phenomenon beyond these current levels can only be explained comprehensively if the current leaking out of the active region is included. Schubert's group identified several leakage mechanisms in blue LEDs comprising QWs in the active region. These include electron–hole recombination outside the QWs due to lack of electron capture into QWs and electron escape from QWs; they offered solutions that include incorporation of electron attraction through a spacer-electron blocking layer by adjusting the p-type doping properties of this layer and minimizing the asymmetry in electron and hole transport properties in GaN.

2.2.5   Blue LEDs' Unique Challenges

Some level of drop in material internal quantum efficiency (IQE) at high currents is seen in all LED structures, including those fabricated in GaN- and GaP-based materials. However, at or near room temperatures, the red LEDs in the GaP/InGaP/AlInGaP material system exhibit a much lower reduction in quantum efficiencies at high currents than blue LEDs in GaN/InGaN/AlInGaN materials do. An example of a slight droop in a GaP-based red LED prototype at 630 nm was demonstrated by Shim et al. (part of Schubert's group) in 2012 and we present their data in Figure 2.6, which shows that the estimated reduction in external quantum efficiency is only 1.6% at 50 mA.

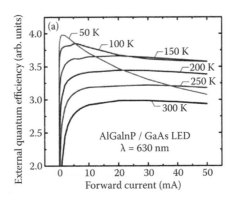

Figure 2.6. Measured external quantum efficiency data versus applied current of an AlGaInP/GaAs LED for different temperatures. The droop magnitude at 50 mA for 300 K is 1.6%, which is significantly less than the droop observed for blue/green LEDs in GaN-based materials under similar conditions. (Reprinted with permission from Shim, J.-I. et al. 2012. *Applied Physics Letters* 100:111106. Copyright 2012, American Institute of Physics.)

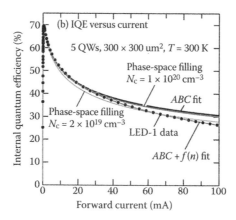

Figure 2.7. Comparison of calculated and measured internal quantum efficiency data for a GaN/InGaN LED as a function of applied current. The droop magnitude at 50 mA for 300 K is 50%, which is significantly larger than the droop observed for red LEDs in GaP-based materials under the similar conditions. (Reprinted with permission from Dai, Q. et al. 2010. *Applied Physics Letters* 97:133507. Copyright 2010, American Institute of Physics.)

Similarly, Osram Opto Semiconductors demonstrated an efficacy reduction of only 3.7% in a GaP-based, 615 nm red LED at 100 mA of injected current [34]. In contrast, the results obtained by Dai et al. (part of Schubert's group), as presented in Figure 2.7, show a reduction in IQE of 50% and 64% at 50 and 100 mA, respectively, for a GaN-based LED device [31]. Such reduction comparison among three different parameters—namely, efficacy, external quantum efficiency, and IQE—is still valid as shown in the following exercise. Note that external quantum efficiency is the same as total efficiency in Equation (1.4) in Chapter 1 when the driver efficiency is 1; for this case, the current is directly applied to the LED chip.

Example 2.1(a)

For a particular monochromatic LED lamp, its efficacy scales linearly with its IQE. This relationship is valid because of the properties of the luminosity function described in Equation (2.10), which will be discussed in Section 2.4, and as long as the LED lamp's light extraction efficiency and driver efficiency remain constant. In practice, these requirements are all satisfied when all the chip and packaging parameters are held constant, which is typical for a particular lamp. Assuming that all chip and packaging parameters remain constant for a particular LED lamp, we can easily show that its efficacy scales linearly with its IQE by using the associative properties of Equation (1.4) in Section 1.5.5 in Chapter 1 as the following:

$$\eta_T = \eta_{\text{int}} \cdot (\eta_{ext} \cdot \eta_{dr}) \qquad (2.4)$$

$$\Rightarrow \eta_T \propto \eta_{\text{int}} \text{ since } (\eta_{ext} \cdot \eta_{dr})$$

is constant and hence, *L.E.* (luminous efficacy) $\propto \eta_T \propto \eta_{int}$. Therefore,

$$L.E. \ (monochromatic \ LED) = k_1 \cdot \eta_T \qquad (2.5)$$

and

$$L.E. \ (monochromatic \ LED) = k_2 \cdot \eta_{int} \qquad (2.6)$$

where k_1 and k_2 are proportionality constants.

It then follows that a certain percentage reduction in monochromatic LED efficacy is equivalent to the same percentage reduction in total efficiency (sometimes referred to as "external quantum efficiency") and internal quantum efficiency.

Example 2.1(b)

Similarly, it can also be argued that for phosphor-based white LEDs, *L.E.* also scales linearly with internal quantum efficiency and total efficiency, as long as the phosphor parameters remain constant. So we can write,

$$L.E. \ (white \ LED) = m_1 \cdot \eta_T \qquad (2.7)$$

and

$$L.E. \ (white \ LED) = m_2 \cdot \eta_{int} \qquad (2.8)$$

where m_1 *and* m_2 are proportionality constants.

Therefore, it follows that a certain percentage reduction in white-LED efficacy equates to the same amount of percentage reduction in total efficiency and internal quantum efficiency.

While the primary reason for pronounced droop in blue LEDs is yet to be determined or broadly agreed upon, it is clear that GaN-based alloy materials currently grown on sapphire and SiC substrates develop a significant amount of defect densities due to the atomic structural and thermal expansion coefficient mismatches between the semiconductor alloy films and substrate materials. This type of hybrid growth is not as ideal as the pure or homoepitaxial growth process, where the epilayers and the substrate belong to the same material system and have the same lattice constants, but may differ only in alloy compositions. Examples of material systems that can utilize this growth process are GaAs and InP, which themselves are used as bulk substrates, and many different alloys with matched lattice constants can be grown on them epitaxially.

Such process can produce high quality single-crystal layers of III–V semiconductor alloys with precise bandgap as well as film thickness controlled within a single molecular layer. These epilayers have very low defect densities and hence exhibit atomically smooth surfaces. Metal-organic chemical vapor deposition (MOCVD) and molecular beam epitaxy (MBE) are suitable techniques for growing these high-quality materials with fine control. MOCVD is widely used for commercial production of compound semiconductor optoelectronic and electronic devices because it is more suitable for large-area crystal growth with higher uniformity. Although LEDs produced in GaAs and InP material systems have exhibited much less droop, they do not emit blue light according to their bandgap energies.

2.2.6   Substrates for Nitride-Based LEDs

Sapphire and SiC thus far still have proven to be the best substrate choices for growing III–V nitride alloys because they have been refined to produce large-area wafers with good crystallinity; further, they are fairly well matched thermally to the AlInGaN system and are compatible with MOCVD approaches. In contrast, bulk GaN—although the ideal substrate for the GaN-based alloy growth in theory—has not yet become a reality because it only comes in very small sizes due to extremely difficult material development processes. Consequently, manufacturers of blue and blue-green shades of LEDs still predominantly employ the heteroepitaxial or hybrid growth techniques and these devices exhibit significantly high defect densities in the LED active region. This is believed to be the reason for lowering quantum efficiencies for light-emitting devices in III–V nitride material systems, particularly at high injected currents.

The droop challenge remains as one of the major bottlenecks against propelling LED lamp industry to a new height. As a result, some LED researchers, engineers, academics, and entrepreneurs are continuing the development of making bulk GaN substrates commercially viable for volume manufacturing of white LEDs [35,36]. They perceive that the large density of dislocations caused by GaN heteroepitaxy is an intrinsic problem for reducing material IQE at high injection currents and that the solution is the development of high-quality GaN substrate. However, no demonstrations have yet been reported on how to synthesize GaN in large, single-crystal form. The inherent difficulties are due to the large refractory properties of GaN (the same property that makes it suitable for wide bandgap device applications), as well as its resistance to hot acids and bases that prevent it from being amenable to any conventional surface preparations that substrates require for epitaxial growth.

As discussed earlier in this section, blue LEDs are used to excite certain phosphors of different colors to generate white light. Most common phosphor used in white LED lamps is yellow, which is generally the cerium-doped yttrium aluminum garnet (Ce^{3+}: YAG). Since white light with different color parameters is desired for various applications, different short-wavelength and monochromatic sources such as blue, violet, and ultraviolet LEDs are now used with various color phosphors to achieve white light with high color rendering index (CRI) and various correlated color temperatures (CCTs) [37]. There are optical losses associated with adding phosphor due to absorption and scattering. These will be discussed in Section 2.4 in relation to increasing overall white LED efficacy.

2.3   Compound Semiconductor Materials and Fabrication Challenges

In contrast to Si-based device manufacturing, III–V compound semiconductor chip manufacturing has lagged in explosive market growth and large price reduction. The lag is more pronounced for general optoelectronic devices in compound semiconductors and even greater for nitride-based, high-brightness LEDs as discussed in the earlier section. Semiconductor properties are inherently sensitive to thermal variations and, in particular, for optical processes. Thus, LED lamps' efficacy, brightness, color quality and stability depend on many parameters and their interdependencies are quite complex. These lighting characteristics vary significantly with chip design dimensions, material quality, and fabrication accuracy.

Among these, the material quality is the most critical factor. It degrades when material morphology and uniformity are poor, which contribute largely to the well-known compound semiconductor yield problem. In the lighting industry, this problem gives birth to "binning" (i.e., the sorting categories that must be created to match LED performances to produce final lamps and luminaires). Significant improvements in high-quality epitaxial growth of compound semiconductors on suitable substrates will be needed to increase the yield for LED lamps substantially.

LED device fabrication involves electrical contact formation, coating, passivation, die bonding and separation, and other processes that make each individual chip suitable for subsequent packaging and testing. All fabricated structures must remain intact to produce stable current–voltage characteristics with minimal leakage current and corresponding light properties over large thermal variations and mechanical stress conditions to ensure reliable lighting performance over the lifetime of the LED lamp. LED engineers need to determine acceptable fabrication tolerances and make sure that the manufacturing platform operates within those tolerance ranges in order to keep LED device performances uniform within certain boundaries.

2.3.1   Yield or Binning Effects for Costs

Current LED lamps that are commercially available to replace incumbent lamps are prohibitively expensive. This high cost will not simply resolve itself with increasing volume. Although for some products, a large and growing market provides a relief for high cost, with factories churning out vast numbers of them efficiently, it does not yet apply to LED lamps because significant complexities exist in manufacturing LED chips with high reproducibility. This is LED lamps' most difficult bottleneck for general lighting applications where color, light level, and distribution parameters must fall within very stringent specifications. The incumbent lamps, although some are not as energy efficient as LED lamps, offer greater degrees of uniformity in light characteristics among a very large pool of products available in the market today. Simply increasing the volume production further to compensate for high costs, without understanding and resolving the complex issues behind the large performance variability that presently occurs in LED chip manufacturing platforms, will only widen the sorting bins. This would pose additional challenges to adopting LED lamps for broad applications. This approach will fail to reduce the LED lamp costs and may even increase them due to more laborious processes of matching significantly more LEDs for targeted applications.

Although the LED industry has grown tremendously, it still only constitutes a very modest segment of the entire lighting industry, which is vast and still growing. But because of LED lamps' unique current and potential advantages, the growth in the LED industry is expected to far exceed the overall growth of the lighting industry. One unique advantage is that LED lamps do not require electric power off the grid. Because approximately 22% of the global population still lives without any artificial lighting and another 14% live with unreliable access to electricity, lamps operating without the need of an electrical grid would perhaps be the quickest and the most practical way of providing light to nearly 2.6 billion people during the dark hours.

However, elsewhere in the world, LED lighting must compete in cost and quality against existing lighting solutions. The long-term savings from energy efficiency argument becomes weak when the initial first cost (IFC) for LED lamps is much higher and the end–end system energy efficiency is still not better or significantly better than that of fluorescent lighting products for most applications. The IFC is a direct result of manufacturing complexities, which must be properly identified, assessed, and resolved. Innovative solutions must be applied all across the supply chain—in particular to improve semiconductor material growth, processing, fabrication, and testing methods to produce reliable, uniform, and reproducible LED lamps.

In order to reduce the IFC substantially, the manufacturing cost savings must occur by improving the current practices in the following five areas:

1. Automated manufacturing

2. Implementation of standards

3. Chip inventory production and management

4. Compound semiconductor technology advancements

5. Large-scale manufacturing process control

Each category is complex and sufficiently unique compared to those in silicon integrated circuit (Si-IC) manufacturing. LEDs' uniqueness stems from a major differentiating characteristic—namely, light! Lighting devices involve significantly more complex engineering designs, simulations, supply chain manufacturing procedures, and, finally, evaluation processes for which a high degree of automation with fast execution times is very difficult to achieve.

2.3.2   Manufacturing Automation

In semiconductor fabrication facilities, Si-IC manufacturing utilizes an astounding level of automation, starting from material formation (wafer growth) to final, packaged modules. This end-to-end reliable automation has led to such inexpensive but sophisticated electronic devices as smart phones and tablet computers. Back in 1965, the power of this type of automation was predicted in "Moore's law" by Gordon E. Moore, the co-founder of Intel, who stated that, in inexpensive integration platforms, the number of transistors in ICs would double approximately every 2 years. Since the 1970s, this prediction has held thus far for Si-ICs. As manufacturing costs further decrease with increasing number of components in Si-ICs over such short periods, the end consumers get to enjoy the low prices of ever more sophisticated electronic gadgets. Despite many predictions and wishes, similar forecasts never came true for the compound semiconductor industry; for the compound semiconductor optoelectronic industry, the cost-reduction scenario has been even bleaker.

Can LED manufacturers borrow automation processes from Si-IC semiconductor manufacturing? Not yet and not entirely. It is clear that LED manufacturers must incorporate more automation and batch processing in the various production stages that test and characterize important optoelectronic and light parameters. Specifically, batch characterizations could be applied for substrates, semiconductor materials, die, and SMD-LEDs. Cost-effective testing and characterizations would entail nondestructive methods (e.g., a system that does not sacrifice wafers and devices when extracting lamp evaluation data). Such automation would require significant investment capital for new testing equipment and methods and for packaging schemes.

2.3.3   Standards Implementation

Implementing standards at every manufacturing step should reduce materials, process, and packaging costs. Standards will also allow fast and effective execution of automated testing. Ideally, LED lamp manufacturing standards would apply at the earliest stages—for example, a lessening of the various substrate sizes and types. However, for white HB-LEDs, the technologies are still undergoing many trial phases and that makes it difficult for settling on substrate standards. For example, in addition to some companies revisiting GaN bulk substrates to overthrow the more common sapphire and SiC substrates, some manufacturers are making notable progress in growing GaN epilayers on Si [38,39]. While bulk GaN substrate would likely improve the epitaxial quality of the diode active material, their sizes would be much smaller than Si substrate sizes. Therefore, any potential cost savings may only be a trade-off for higher quality.

To overcome such trade-offs, other groups believe in first growing high-quality GaN-based material in best-suited smaller substrates (typically 2 or 4 in. diameter) and then applying epitaxial lift-off to bigger wafers for subsequent fabrication steps [40]. Such hybrid processes could be the optimized low-cost solution because higher throughputs via an 8 in. substrate manufacturing line would generate significantly more LED devices per run.

Such dissimilar solutions pose corresponding challenges for other manufacturing stages where frequent ideas, changes, and innovations complicate the standard adoption for device design layouts, fabrication procedures, and testing methods. Still, the LED industry must focus on implementing such standards because it should also boost interoperability among various LED supply chain vendors, which can reduce IFC through legitimate industry competition and speed up new product entries to the market.

2.3.4   Chip Production and Inventory Management

The LED industry has two segments: HB-LEDs, with their many variations, and the rest. All HB-LEDs eventually move to various applications, ranging from display back illumination, refrigerator lighting, retail, street, automotive, and general lighting. Many manufacturers do not have systematic production lines or inventory management processes that churn out different LED lamps for these applications. Instead, the current selection process includes manual testing and matching of many HB-LEDs, which is both costly and time consuming. Ideally, perhaps, manufacturers would produce only a few types of LED modules generated from correspondingly few production lines and differentiated only by one or two parameters, such as their light output level, efficacy, or color temperatures.

Further, if manufacturers placed color phosphors in the secondary optics portion of the package instead of coating them on the individual LED chip or module encapsulation, the color temperature differentiation could be pushed closer to the final packaging stage. Such adoptions would make manufacturing processes more flexible and scalable, leading to faster turnaround times with better inventory management. An example of such a configuration is shown in Figure 2.8, which

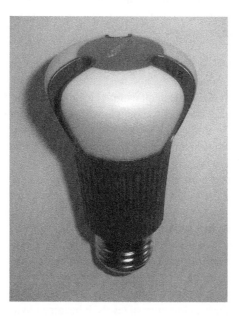

Figure 2.8. A 12.5 W LED lamp from Philips that replaces a regular A-19 60 W incandescent lamp. The crown-shaped dome is coated with yellow phosphor to generate white light from blue LED chips placed inside the lamp.

is an A-line LED bulb from Philips that uses phosphor in the outermost surface of the lamp.

However, phosphor separation from the LED chips would require obtaining and applying the appropriate set of lighting parameters for different applications, which can be adopted using Table 1.3 in Chapter 1. The advantage of this method would be that the same HB-LED chips or SMDs can be used for a variety of applications because adjusting the color as well as light distribution characteristics can now be incorporated in the final stages of the production, rather than at the front end.

2.3.5   Technology Advancements of Compound Semiconductors

As discussed previously, the complexity challenge exists in generating light from compound semiconductor materials reproducibly because of the numerous physical parameters' interrelations and significant temperature dependencies. Extensive theoretical physics and practical engineering developments of compound semiconductor materials and processing are needed to achieve fine control of these parameters in manufacturing environments and, thereby, improve LED light quality, efficiency, and product yield.

Compound semiconductors, GaN-based materials in particular, have higher defect densities and consequently require extensive engineering and stress management processes for improving LEDs' efficacy and reliability. These are still under investigation and therefore—unlike Si-IC manufacturing that already has developed reliable links among its actual materials, processes, and corresponding simulations in a large-scale manufacturing platform—the GaN-LED has yet to develop such an infrastructure.

2.3.5.1   LED Materials Science

The development of materials science for LEDs has been remarkable since the first blue LED in AlInGaN was demonstrated by Nakamura at Nichia. Early such LEDs degraded quickly over time and operated based on donor–acceptor optical recombination, which produced a rather broad spectrum that works against providing color saturation. For a more stable LED with good color saturation, radiative recombination should only result from direct band-to-band transitions. Early Nichia LEDs also had a small AlInGaN alloy composition range that offered limited means for lattice matching between active layer and substrate materials. Broad quaternary alloy composition range is helpful for minimizing dislocations introduced by lattice mismatches.

The role of dislocations in optical recombination processes in the III–nitride alloys is an enduring research topic among academics and the LED industry. Semiconductor scientists have been surprised because, despite the large dislocation densities in these materials, light emission from blue LEDs is appreciable nevertheless, particularly at low current levels. It became obvious that dislocations are less harmful to nitride devices compared to arsenide or phosphide devices with corresponding physical properties. However, scientists still believe that dislocations weaken device efficiencies. They may impair operation by increasing nonradiative recombination at or near dislocations, trapping unwanted impurities during the growth process, and impeding carrier velocities.

Substantial reduction of dislocations in nitride-based materials may be achieved with growth strategies that incorporate (1) effective buffer layers that allow the deposition of a "seed" nitride layer free of dislocations, (2) lattice matching of all subsequent epitaxial layers with a single lattice constant to the "seed" nitride layer by means of quaternary layer structures, and (3) appropriate substrate choices. Proper investigation through characterization of the crystallographic, electronic, and light properties of the alloys will enable successful demonstration of single-crystal GaN-based materials with low dislocation densities.

Another degree of difficulty arises for producing visible LEDs in the entire blue-green portions of the spectrum using only band-to-band transitions. This process demands fabricating low-stress films with low-bandgap energies. Although low-bandgap GaAsN alloys can produce emission in the wider green spectrum, the LED industry has chosen large-bandgap InGaN materials due to the high bond strength and better color saturation. The latter requires staying justified to a critical epilayer thickness in InGaN in order to maintain low dislocation densities while employing only band-to-band recombination. Such optimization can be achieved by developing theoretical band structure models to study the material characteristics that include band alignment, doping properties, and transition strengths.

2.3.6 Large-Scale Manufacturing Process Control

If the core reasons for wide performance variability in GaN or other compound semiconductor-based LED chips are not resolved, large-scale manufacturing automation, process control, and standardization will remain difficult. Engineers can only establish reliable transitions between subsequent processing stages if all variations or failures can be quantified, tolerated, and, in turn, appropriate parameters can be adjusted to continue in-line processing in order to produce acceptable characteristics for final throughputs. Creating such a platform for LEDs would require establishing process links among material databases, process simulation, complete device simulation, and package simulation that incorporates optical, thermal, mechanical, and electrical aspects. These will be needed to mirror the foundations of the Si-IC semiconductor process control domain, which include run-to-run control, data mining, and equipment tracking.

Without such process-control manufacturing platforms that are undeniably linked to the compound semiconductor material growth and fabrication challenges, the benefits of larger diameter substrates will be few. Similarly, benefits from larger chip sizes, as well as batch tooling of material characterization equipment, will not significantly affect cost reductions for LED lamps. While the challenges for building well functioning, large-scale LED manufacturing platforms may seem insurmountable, the solid-state lighting industry has already begun to acknowledge the unique difficulties of producing low-cost, high-quality LED lamps. When the manufacturing challenges are within fine control, the final achievement or outcome will present itself in higher binning yield and high-quality LED replacement lamps with much lower IFCs.

2.4   Determining and Improving LED Lighting Efficacy

In the preceding chapter and sections we have seen that monochromatic LED lighting efficiency depends on internal quantum efficiency, light extraction efficiency, and driver efficiency as given by Equation (1.4) in Section 1.5.5. Prior sections included discussions on how material IQE is affected by substrate, materials, and fabrication parameters because of dislocation densities, transition types, electrical contact qualities, and others. In this section, we shall see how the total efficiency can be quantified and how to estimate LED efficacies for both monochromatic and white light. We shall also discuss how to improve various light extraction efficiency factors and then attempt to estimate the practical efficacy limit ranges for common white HB-LEDs.

2.4.1   Quantification of LED Efficiency and Efficacy

An LED produces monochromatic light when forward biased and its color is determined by the bandgap energy of the semiconductor material based on the relation given in Equation (2.3). As discussed in the previous sections, the light emission efficiency largely depends on active material qualities. Because of the challenges associated in group III-nitride growth, LEDs in the blue and blue-green part of the spectrum have much lower material IQEs than those from the red spectral regions that are constructed with GaP/InGaP/AlInGaP material systems.

High-power blue LEDs that use AlInGaN active material have IQE near 70%, whereas red LED counterparts comprising the AlInGaP active region have IQEs of almost 100% [41]. The total internal quantum efficiency is the product of material IQE and device efficiency. The device efficiency depends on the applied electrical drive current's ability to affect radiative recombination in the active region. Consequently, it depends on the active layer design that often includes QWs, as well as the electrode configurations and contact resistances. In order to maximize device efficiency, the voltage drop across the p–n junction should be uniform and the entire electric field strength should overlap the whole junction region without any spillage outside the active region. In the active region, all carriers should be confined and be allowed to recombine radiatively with their oppositely charged counterparts.

The amount of light coming out of the LED module is limited by the light extraction efficiency, η_{ext}, which depends on internal absorption by materials, as well as reflection and scattering from various surfaces and materials in the system, including phosphor for phosphor-based LEDs. Appropriate index matching and high-quality phosphor and semiconductor materials will improve the overlap integral of optical power transfer from the chip-encapsulation ensemble to any external region. Finally, the efficiency of the driver, η_{dr}, which delivers the injected current into the semiconductor diode, limits the overall η_T.

Therefore, to include all terms, Equation (1.4) can be rewritten as

$$\eta_T = (\eta_{IQE} \cdot \eta_{dv}) \cdot (\eta_{Lop} \cdot \eta_{Pop}) \cdot \eta_{dr} \qquad (2.9)$$

where η_{IQE} and η_{dv} are material IQE and device efficiencies, respectively, and η_{Lop} and η_{Pop} are optical throughput efficiencies of the LED and phosphorus respectively.

Phosphor-based white LEDs experience some efficiency degradation due to the thermal loss associated with the Stokes shift as well as light scattering losses. These may be minimized by optimizing the choice of phosphors, their location in the module package, and their physical parameters and deposition techniques. Increasing the luminous efficacy of phosphor-based white LEDs will require further achievements in phosphor loss minimization as well as developing more efficient short-wavelength LED light sources.

Conducting the efficiency and efficacy measurements and calculations is more straightforward for monochromatic LEDs than for white LEDs. The experimental determination of total energy efficiency, η_T, for monochromatic LEDs is simply done by measuring the radiant power in watts and dividing that by the electrical power in watts applied to the LED. The luminous efficacy of these monochromatic LEDs then is determined from first calculating the luminous flux using

$$F_{m\text{-}LED} = (683.002 \text{ lm/W}) \cdot \int_{L_1}^{L_2} V(\lambda) \cdot J(\lambda) d\lambda \qquad (2.10)$$

where $V(\lambda)$ is the dimensionless, standard CIE photopic luminosity function (typically, $L_1 = 380$ nm and $L_2 = 780$ nm), and $J(\lambda)$ is the spectral power distribution of the optical radiation in watts per meter. Dividing $F_{m\text{-}LED}$ by the electrical power in watts applied to the LED, the luminous efficacy is then easily obtained.

If the LED's dominant source wavelength is at 555 nm and the color is saturated with all the optical power concentrated at this wavelength, then luminous efficacy is the product of η_T and 683. This, of course, is an ideal case. However, a crude estimation of efficacy can be made of some of the best nearly saturated R&D green LEDs that have exhibited η_T of nearly 60% at 555 nm, which would be approximately 410 lm/W. (In the case of a 555 nm green laser, this approximation is more valid.) However, for the same efficiency, the luminous efficacies for blue LEDs are much smaller, which can only be accurately estimated by applying Equation (2.10). Figure 2.9 shows how the application of Equation (2.10) leads to the total lumen output of a blue LED.

Blue LEDs are most commonly used to produce white LEDs and therefore it is important to determine how efficient blue LED chips are, how much more they can be improved, where the improvements need to occur, and, finally, what the corresponding white LED's practical efficacy limit is.

In order to determine the efficiency factors for a blue LED, Equation (2.9) should be modified omitting the phosphor efficiency factor as

$$\eta_T = (\eta_{IQE} \cdot \eta_{dv}) \cdot (\eta_{Lop} \cdot \eta_{dr}) \qquad (2.11)$$

Determining the efficiency of a blue LED in this manner is particularly helpful for white LED lamps that use phosphors in the secondary optical elements.

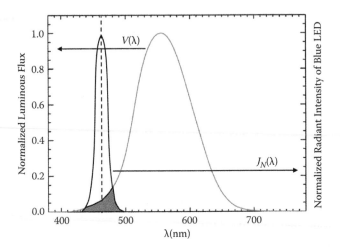

Figure 2.9. A measured InGaN blue LED's normalized spectral power distribution, $J_N(\lambda)$, is plotted with the CIE $V(\lambda)$ photopic luminosity function against wavelength in nanometers. The shaded area is the overlap between the two functions, which gives the LED luminous flux when multiplied by the appropriate proportionality constant.

Example 2.2

Let us now estimate some values for the various efficiency factors in Equation (2.11) using some of the best reported efficiency values for commercial production-quality blue LEDs and assuming certain known, typical values for some of the factors in the equation. Thus, we gather the following parameters for Equation (2.11):

η_T = 0.58 (typical high-end values for commercial, high-power blue LEDs)

η_{IQE} = 0.7 (Dai et al., Figure 2.7; represents calculated or measured values of material IQE)

η_{dv} = **x** (includes droop arising from all device efficiency degradation factors)

η_{Lop} = 0.95 (assumption made using best industry values for light extraction from the LED chip; degradation includes device optical reflection, internal scattering, and absorption)

η_{dr} = 1 (the blue LED chip is directly driven with a current source)

Applying Equation (2.11), one gets

$$\eta_{dv} = 0.87$$

This means that the droop degradation is approximately 13%. This figure is encouraging because, just a few years ago, the droop degradation was quantified to be around 55% at current levels exceeding 100 mA.

The industry has reported corresponding typical efficacy values of 125 lm/W when blue LEDs of the type in Example 2.2 are used. This claim is also supported in the announcement made by Osram Opto Semiconductors in February 2012, which reported some encouraging research results from GaN-based LEDs fabricated on 6 in. silicon substrates [39]. They succeeded in manufacturing these GaN-on-Si-LED devices in their standard Golden Dragon Plus packages, which produced 634 mW at 438 nm dominant wavelength, for a drive electrical input power of 1.102 W—yielding an efficiency of 57.5% at the chip level. These results are on a par with their GaN-LEDs on sapphire substrates. Osram stated that their projected low-cost GaN-on-Si process could be ready for production within 2 years, eventually increasing the Si substrate sizes for lowering costs even further.

Osram also reported a luminous efficacy of 127 lm/W from a packaged white LED that used these blue LEDs and conventional yellow YAG phosphor, with a CCT of 4500 K. They stated that the efficacy was determined from an obtained value of 140 lm for luminous flux at injected current of 350 mA and $V_f = 3.15$ V.

The luminous flux of a white LED such as those manufactured by Osram in the preceding discussion is determined by measuring its spectral intensity or power distribution over the visible region and overlapping it with the $V(\lambda)$ function as follows:

$$F_{w\text{-}LED} = (683.002 \text{ lm/W}) \int_{L_1}^{L_2} V(\lambda) \cdot K(\lambda) d\lambda \qquad (2.12)$$

where $V(\lambda)$ is the dimensionless, standard CIE photopic luminosity function, L_1 an L_2 are the same as in Equation (2.10), and $K(\lambda)$ is the spectral power distribution of the optical radiation in watts per meter. Dividing $F_{w\text{-}LED}$ by the electrical power in watts applied to the LED, the luminous efficacy is then easily obtained.

Figure 2.10 shows the normalized values of a measured $K(\lambda)$ function plotted along with $V(\lambda)$ function over the same wavelength abscissa. This allows the numeric computation of the light output in lumens (i.e., $F_{w\text{-}LED}$, using Equation 2.12).

2.4.2 Determination of Phosphor Efficiency

Phosphor efficiency can be experimentally determined by measuring $J(\lambda)$ and $K(\lambda)$ of Equation (2.10) and Equation (2.12) respectively. Then, integrating the total power contained in $K(\lambda)$ and dividing it by the total power contained in $J(\lambda)$ will yield the phosphor efficiency. Thus, the phosphor efficiency is given by

$$\eta_{Pop} = \frac{\int_0^\infty K_N(\lambda) d\lambda}{\int_0^\infty J_N(\lambda) d\lambda} \qquad (2.13)$$

where $K_N(\lambda)$ and $J_N(\lambda)$ are normalized functions of $K(\lambda)$ and $J(\lambda)$ respectively.

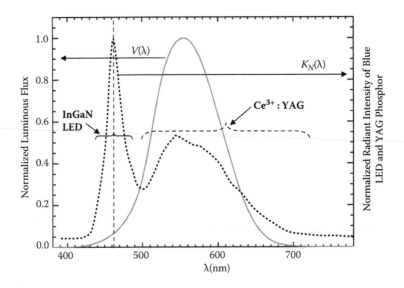

Figure 2.10. A white LED's measured normalized spectral power distribution, $K_N(\lambda)$, is plotted with the CIE $V(\lambda)$ photopic luminosity function against wavelength in nanometers. The overlap between the two functions gives the LED luminous flux when multiplied by the appropriate proportionality constant. The white LED in this example is composed of a standard InGaN LED and common Ce^{3+}:YAG phosphor.

Equation (2.13) yields the total phosphor efficiency that is the product of various constituent efficiency factors related to thermal and scattering losses. In order to determine the effects of these factors and to quantify them, various physical parameters may be changed strategically and η_{Pop} may be recalculated from corresponding $K(\lambda)$ spectra. An empirical optimization process in conjunction with some simulations of light throughput in different phosphor materials with varying parameters should establish a method for minimizing phosphor losses. The process will establish the optimum choices of phosphors, their location in the module package, deposition techniques, and their physical parameters such as thickness, density, and uniformity.

2.4.3   Estimation of White LED Efficacy Limit

Let us now see what maximum efficacy is expected as a theoretical limit, as well as a practical limit, for phosphor-based white LEDs that make use of blue LEDs as discussed in this section. In Chapter 1, we discussed the trade-off relationship between efficacy and CRI in white LEDs—meaning that efficacy can be increased by reducing CRI. To best avoid the effect of this trade-off, we will only look at maximizing the efficiency factors for white LEDs that use the same type of blue LEDs and conventional yellow phosphors, so as to keep the CCT and the CRI nearly constant. For this confined system, the efficiency factors that can be further improved are η_{IQE}, η_{dv}, and η_{Lop}, and η_{Pop}.

2.4.3.1 Theoretical Efficacy Limit

We have seen in Section 2.4.1 that, according to some of the industry's best achievements, including that from the Osram LED, a total chip efficiency of 58% from a blue LED leads to an efficacy of 127 lm/W for a white LED packaged conventionally with this blue LED and common yellow phosphor. Making use of *this* result and Equation (2.7), we establish the following:

If, for,

$$\eta_T = 0.58, L.E. \approx 127 \text{ lm/W}, \tag{2.14}$$

it follows that

$$\text{for } \eta_T = 1.00, L.E. \approx 219 \text{ lm/W}. \tag{2.15}$$

Efficacy of 219 lm/W would be the theoretical limit for a conventional phosphor-based white LED, if phosphor losses were zero. However, because this is not likely and phosphor losses are not accurately reported in the industry for these white LEDs, we assume a reasonable phosphor degradation of 10% in this case.

Therefore, assuming that the phosphor loss may have reduced the efficacy by 10%, adding this 10% should lead to the following maximum limit: Since

$$(90\% \text{ of } L.E._{\text{max-theoretical}}) \approx 219 \text{ lm/W}, \tag{2.16}$$

it follows that,

$$L.E._{\text{max-theoretical}} \approx \frac{219}{0.9} \text{ lm/W} \approx 243 \text{ lm/W}. \tag{2.17}$$

Hence, maximum theoretical efficacy limit for a white LED packaged with common yellow phosphor is then estimated to be 243 lm/W.

Note that there are several assumptions made here that may not be entirely correct, such as the phosphor loss quantity and phosphor loss linearly degrading the efficacy. In practice, improving the phosphor loss may mean changing $K(\lambda)$ distribution, which would affect $L.E.$ in a nonlinear fashion. This would also imply that CRI and CCT may not remain constant. However, it is expected that if one works with the same phosphor material, these effects would be small.

2.4.3.2 Practical Efficacy Limit

While the theoretical efficacy limit is useful to know as a ceiling, the knowledge of a credible practical limit for the widely used white HB-LEDs is also beneficial for the industry. To arrive at such a number, let us consider some targets according to "realistic" (although not proven) future improvements of the individual efficiency factors. These are assembled for a typical blue LED chip as follows:

η_{IQE} = 0.8 (current number is 0.7 from Dai et al., Figure 2.7)

η_{dv} = 0.95 (includes droop arising from all device efficiency degradation factors)

η_{Lop} = 0.96 (assumption made using best industry methods for light extraction from the LED chip; degradation includes device optical reflection, internal scattering, and absorption)

η_{dr} = 1 (the blue LED chip is directly driven with a current source)

Applying Equation (2.11), one gets

$$\eta_T = 0.73.$$

Using Equation (2.14) and Equation (2.7), we arrive at an efficacy limit of

$$L.E._{practical} \approx 160 \text{ lm/W}, \tag{2.18}$$

for $\eta_T = 0.73$ and for the same phosphor degradation of 10% as applied to Equation (2.16). Further, if we assume a practical, industry-best electrical driver efficiency of 0.95, $L.E._{practical}$ is further reduced to

$$L.E._{max\text{-}practical} \approx (160 \times 0.95) \text{ lm/W} \approx 152 \text{ lm/W}. \tag{2.19}$$

This maximum practical efficacy assumes that IQE of nitride-based materials is further improved, the droop and other device inefficiencies are substantially reduced, and a great majority of the generated light is extracted from the packaged LED. Although these are proving to be very difficult to achieve, particularly in manufacturing environments, these are the industry goals that may be attainable by means of breakthrough technologies in LED materials, devices, packaging, and electronics. Such achievements will invariably include design enhancements and rigorous engineering as well as manufacturing optimizations to maximize all the efficiency factors.

Although Equation (2.17) and Equation (2.19) offer theoretical and practical efficacy limits for the most conventional white LEDs in today's market, it is important to note, however, that white LED efficacy may be further enhanced by utilizing green or other types of monochromatic LEDs along with perhaps violet or other lower loss phosphors. These are now under development at various companies and research laboratories.

③ LED Module Manufacturing

3.1 Introduction

In Chapter 2 we have seen that LED light output characteristics depend on many parameters, most of which stem from one of the most significant natural behaviors of semiconductors; that is, their optical properties are inherently very sensitive to thermal variations. Strictly speaking, the static semiconductor optical and optoelectronic property dependencies on temperature are nonlinear *and* they also vary nonlinearly with time [42], which invariably complicates an LED lamp's performance over its life span. The solid-state lighting (SSL) industry's goal must be to design LED modules by optimizing a set of parameters that will produce reasonably stable light over long periods of time with competitive—but, more importantly—advantageous illumination performances compared to those from existing lamps while ensuring that their manufacturability aspects remain straightforward.

In this chapter, the constituents of an LED module, or light engine, will be discussed within the context of a subsystem that encompasses several engineering domains; the temperature dependencies will be investigated in Section 3.3 to help design light engines with more effective thermal management in order to increase LED lamp life span. The final section of the chapter will elaborate on how an LED engineer would pursue optimization of module parameters that will produce first-rate engines and luminaires without increasing manufacturing complexity to enable product realization within the means of low-cost factory environments.

3.2   LED Lighting Components and Subsystems

In contrast to incandescent and fluorescent lamp configurations, LED lamps have many more parts and elaborate configurations that are constructed using intricate fabrication procedures such as those discussed in Chapter 2. Invariably, LED lamp's efficacy, brightness, color quality, stability, and longevity depend on a large number of parameters associated with these components and manufacturing techniques. Because many parameter interdependencies are also quite complex, the lighting characteristics vary acutely with chip design dimensions, material qualities, fabrication variability, and packaging methods. Production of high-quality LED lamps with high drive power therefore requires rigorous application of thermal, optical, electrical, and mechanical engineering disciplines and verification methods. Consequently, an LED lamp is not merely a small constituent of a luminaire, but rather a subsystem incorporating several physical domains. For some applications, such a subsystem may require refined sensing and control of thermal, electrical, and light parameters during operation in order to enhance LED luminaires' performance and life span.

3.2.1   Power-Based LED Module Configurations

Initial introduction of LEDs as indicator lamps found popular usage primarily in circuit boards. They only require several 10s of milliamps of input current, limiting drive power in the range from a few milliwatts to no more than a few 10ths of a watt. These lamp categories typically include 5 mm (T1-3/4) and T1-type module structures with other diameters, which have become widely recognized single-color and white LED lamps. An assortment of these lamps is shown in Figure 3.1.

Such low-power lamps typically emitting only a few lumens generate little thermal power and therefore do not require any additional physical structure to remove the corresponding heat from the diode junction area. The 5 mm, T1-3/4 design is inadequate for handling electrical drive power of 1 W or more, which has become the typical wattage for a single high-power, high-brightness (HB)-LED module [43].

High-power HB-LED module design, development, and manufacturing techniques have significantly improved in the past several years; this has led to numerous sophisticated, commercial engines and subsystems with remarkable performances and features from an increased number of suppliers. Such an engine or a single, mountable LED module in its simplest form uses a housing that includes a heat sink at the base of the case, on which one or several LED chips are directly mounted as was shown in Chapter 2, Figures 2.4(a) and 2.4(b). Here, a cross-section of this basic module with external connections is shown in Figure 3.2 to emphasize its usability in the next level of the supply or manufacturing chain. These modules are attached to certain printed circuit boards (PCBs) and often to additional thermal management structures, which enable capacities of more than 1 W of drive power that can generate well over 100 lumens per LED engine over a long life span, even when they contain only a single LED chip.

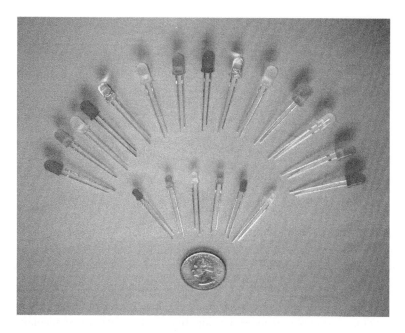

Figure 3.1. An assortment of widely recognized 5 mm T1-3/4 LED modules (top row) and 3 mm T1 LED modules (bottom row) are shown along with a US quarter for size reference. These emit a few lumens, typically requiring a few milliwatts of drive power.

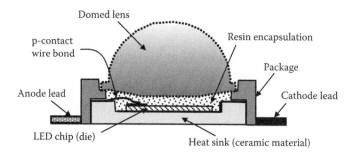

Figure 3.2. Schematic cross-section of a basic high-power LED module or engine that can be soldered on a PCB or placed in a luminaire housing.

The elements of a module type shown in Figure 3.2 are primarily conducive to two main goals: (1) effective heat dissipation, and (2) surface mountability on PCBs. Accomplishing the second goal should also favor the first—that is, how the module is mounted should also maximize its heat dissipation capacity. Thus, it is recommended that the PCBs contain metallic regions so that power LEDs can be directly soldered to the metal in order to ensure effective heat transfer from the LED heat sink to the PCB.

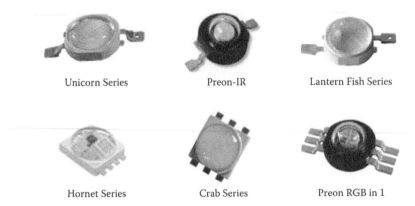

Unicorn Series   Preon-IR   Lantern Fish Series

Hornet Series   Crab Series   Preon RGB in 1

Figure 3.3. A variety of high-power LED module from ProLight Opto, a manufacturer in Taiwan. These modules may be integrated on a PCB or may be directly placed in a luminaire system with additional thermal management equipment and driver electronics. (Photo courtesy of ProLight Opto.)

For white LEDs, phosphor coating is either applied directly on the chip or on the lens or enclosure placed above the chip. In order to protect the semiconductor chip, an encapsulation is formed on it with a resin coat. Manufacturers now have the choice of mixing phosphor with this resin as well. Phosphor and resin coatings or their mixture may be applied using a single fluid dispensing system [44]. It is important that these coatings be optically homogeneous over the module's entire life span. Further, their dispensed volume and consistency must be uniform and repeatable in order to maintain module performances of manufactured units within tightly bound ranges to generate high yield or throughputs.

Many variations of the package shown in Figure 3.2 are now available in the market to address different applications and various levels of integration to construct complete luminaires. An ensemble of such modules is shown in Figure 3.3, all of which are from the same manufacturer.

Numerous other module varieties are also available in the market, often making it difficult for luminaire manufacturers to settle on their designs at the higher supply chain levels. While many module varieties may offer different design opportunities for niche markets, standards for form factors and performance specifications would be necessary for high-volume lamp and luminaire markets. Only with such appropriate standards met by a reasonably small number LED module or engine footprints with closely matching performances would the manufacturers achieve interoperability among several types of lamp or luminaire structures.

3.2.2   LED Subsystem Configurations

Just as we have noted the many module variations in the previous section, there is even a wider variety of subsystem configurations in the current market. This is reminiscent of the SSL industry still experiencing a maturing phase. At present, the many physical and functional varieties pose a difficulty for constructing a bound definition of a subsystem. At a high level, we may define a "subsystem" as a central part of an LED luminaire: *An LED subsystem is a construct that includes a number of functions and controls that are integrated on a platform by means of an appropriate arrangement of several elements or components, including one or more LED chips or packaged emitters, in order to generate the main output (i.e., illumination), which can be clearly recognized, appreciated, and utilized by the end user.* Such subsystems include integration technologies similar to electronics, which often incorporate a chip-on-board (COB) platform for combining electronic circuitry on a global PCB, along with a different underlying core substrate of higher thermal conductivity to serve as a global heat sink. This heat sink helps to remove heat from the LED modules (or encapsulated chips on submounts, which are often referred to as "emitters") and other such active components as driver integrated circuits (ICs) and allows thermal dissipation into the surrounding environment.

As discussed previously, thermal management is crucial for LED lamps and luminaires. It remains as a central issue for the module, subsystem, and, finally, the entire luminaire construction since the high heat generated in the diode junction region must be continuously removed during operation by means of conduction and convection. Luminaires that house incandescent and various fluorescent lamps are unsuitable for providing the thermal management functions necessary for LEDs. LED luminaires require many of the technologies developed and utilized for cooling semiconductor devices such as computer processor chips and lasers.

The SSL industry has been benefiting from the other electronic and optoelectronic industries that preceded it. In particular, the semiconductor laser and transponder industries supporting the Internet and fiber-optic telecommunication network equipment have paved the way for high-power LED lamp design and manufacturing. The vast set of proven technologies used in the previous optoelectronic technologies relating to thermal and mechanical aspects along with incorporation of electronic integration on boards have allowed sophisticated subsystem technology development for LED luminaires within a fairly short period of time. Thus, LED engineers and manufacturers have quickly adopted various PCB and COB technologies to produce various levels of thermal management, optical output, and electronic controls.

3.2.2.1 Printed Circuit Board Subsystems

The selection of PCB material and design largely depends on what magnitude and distribution of optical intensity or flux are required from certain desired luminaire-housing size and shape, in order to determine how much and what methods of cooling would be necessary. The three most common PCB types for related industries are FR-4 (flame resistant 4), metal-core PCB (MCPCB), and ceramic PCB. Although most common for low- to moderate-speed electronics, FR-4 is generally not suitable for ultra high-power LEDs, which require higher thermally conductive baseboard materials. In FR-4 PCBs, localized metallic regions are embedded and the dielectric base material is FR-4, which has thermal conductivity of only 0.25 watts/millikelvin (W/mK) in contrast to 300 W/mK and 150 W/mK for copper and aluminum alloy respectively [44]. In some cases, however, several manufacturers have produced luminaires with power LEDs by carefully designing FR-4 PCBs with thermal vias [45,46].

MCPCBs, also known as insulated metal substrates (IMSs), use a much higher thermally conductive dielectric material, which acts as a thermal bridge between the LED chip and the metal back plate, while providing electrical isolation for IC components. Most luminaire manufacturers utilize MCPCBs for the majority of the SSL applications; however, some use ceramic PCBs for most demanding uses such as in aerospace or other industries that require reliability in high-pressure and -temperature environments.

Heat dissipation from the heat sink embedded in the LED module (or emitter) to the PCB may be maximized by directly soldering power LED modules to the metal core of MCPCBs. The advantage of an MCPCB is that it also removes the heat from the electronic driver ICs often surface mounted on the PCB. Luminaire engineers must carefully design the overall size of the PCB and the spacing of LEDs, as well as other ICs, so that adequate heat removal can be achieved to keep the LED die temperature under a certain level. For many high-performance LED luminaires, a large single extruded heat sink is usually bonded to the bottom of the PCB using high-quality thermal compound, as shown in Figure 3.4. In order to further enhance thermal management, many decently sized holes may be formed on the luminaire housing to create air flow and heat dissipation by means of convection. In some cases, the lamp or luminaire housing itself could serve as additional heat sink, as shown in Figure 3.5.

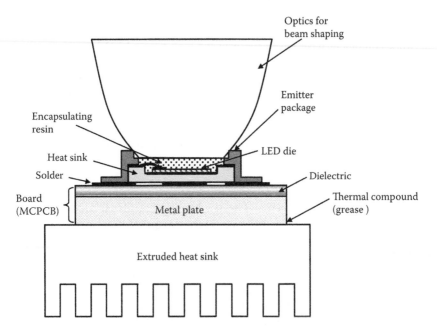

Figure 3.4. Schematic drawing (not to scale) of an entire LED subsystem construct of a luminaire showing external optics and an LED emitter mounted to a PCB, which is bonded to an extruded heat sink.

Figure 3.5. A photograph of a dimmable PAR20 LED lamp showing the extruded heat sink as part of the housing. It produces 350 lm and consumes only 8 W of electrical input power.

3.2.2.2 Chip-on-Board Subsystems

As the IC industries develop newer and more sophisticated technologies, manufacturers incorporate higher degrees of integration in subsystems, which in turn produce slimmer gadgets with more capability. The SSL industry is no exception in this regard. Several manufacturers are already moving forward with integrating LED chips directly on PCBs—that is, utilizing COB technology [47,48]. This approach bypasses the production steps of surface mount device (SMD) modules as well as some PCB integration levels and thereby reduces overall manufacturing complexity for LED luminaires. As the technology matures, this may ultimately result in lower costs, higher performance and functionality from slimmer packages, and better thermal management configurations.

Another advantage with this technology is that it allows for a larger ratio of emitting area to the overall packaged subsystem because LED chips can be soldered directly onto the board with higher density. This will improve luminance uniformity for the luminaire—a characteristic that is desired for various illumination lamps that will be discussed in the upcoming chapters of this book. Denser arrangement of chips will also enable smaller pixel pitches in LED-based electronic message centers (EMCs)—a necessary feature for on-premise EMCs that are to be viewed from close distances. In addition to the maturation still needed for the COB technology, its application will likely be limited to retail, down-, and high-bay types of lighting because illumination can only be generated from flat surfaces, unless a suitable flexible substrate can be adopted. In contrast, if modules are used, they allow the options to be mounted on many slanted or mutually orthogonal surfaces and therefore can produce light at different angles.

Luminaires using the COB technology will also need increased thermal management to accommodate for higher emitting area ratios *if* more heat is generated per square area from the entire luminaire housing, which is likely.

3.2.2.3 Electronic Drivers

As discussed in Chapter 2, an LED only emits light when the p–n junction is forward biased and a net diffusion current flows from the p-region to the n-region, allowing many minority carriers to recombine radiatively in the junction. The current–voltage relationship of an LED is reasonably characterized by the *Shockley ideal diode law* [49], which describes that the LED current, I_F, rises exponentially when a positive voltage V_F across it exceeds V_{bi}. This implies that a small change in voltage above V_{bi} leads to an exponential change in current. Further, a voltage input drive will generate unpredictable light output for different LEDs due to manufacturing variations in V_{bi}, which is further complicated by its inherent temperature dependence. Thus, driving an LED with a constant voltage source is not recommended because it could potentially destroy the device if the maximum voltage rating is exceeded even by a slight amount. More prudent is to use a constant current source as the LED driver while ensuring the voltage drop across the diode remains well under the maximum rating. Since the LED light flux output, L (previously also denoted as Φ), as a function of I_F is fairly linear in the range where V_F is safe, a constant current driver also provides a very good control or predictability of L against I_F, when I_F serves as the driver input quantity.

Most common power supplies provide a constant voltage output such as batteries and mains and therefore an LED driver requires an additional power converter to generate a constant current source. Such a converter uses I_F as the error signal in a feedback loop to alter the voltage from the power supply in order to maintain a constant I_F. Since direct measurement of I_F is often difficult, monitoring other parameters (e.g., a sensor resistor value within the corresponding circuit) can allow for a good estimate of I_F.

In order to support the wide variety of modules and subsystems that are finding increasingly many lighting applications around the world, different driver types are also proliferating in the SSL industry. As the industry boasts in providing highly energy-efficient lighting, it is very important to ensure that the usage of inappropriate drivers does not diminish the wall plug efficiency significantly. If one category of LED luminaires varies significantly in terms of input power and voltage requirements from another category, using the same driver under the same conditions would yield different power efficiency, power factor correction (PFC), and total harmonic distortion (THD) for the two groups. Since it is practical to keep the driver categories to a reasonable number, drivers that can select the appropriate PFCs and minimize THDs for groups of LED lamps and lighting systems will be very desirable.

Primarily, there are two basic types of driver functions: nondimming and dimming. Because LED light color properties shift when intensity is modified, nondimmed LED lamps usually provide the most reliable and high-quality lighting performance. However, it relinquishes energy saving opportunities during times when dimmed lighting would be sufficient. Ultimately, overall trade-offs involving cost, performance, and perhaps other application-specific requirements should determine the right choice of LED drivers.

3.2.2.3.1 Nondimmable LED Drivers Various nondimmable driver configurations are necessary to support different power ranges (e.g., 1–25 W) as well as voltages of mains (i.e., 100–120 V and 230 V) and rails (e.g., 6 V for emergency lighting, 12 V AC/DC in home lighting, and 24/48 V for street lighting). These varieties, along with lamp configurations that may yield different light intensity and distributions when several combinations of LEDs in series and parallel are used, can further complicate the idea of a single driver providing optimum solutions for all intended applications. However, a family of nondimmable drivers to suit groups of like voltage and power ranges could be developed so as to keep the proliferation under control. Without adding further varieties with dimmable functions, it suffices to say that, considering many trade-offs, simpler nondimmable drivers may be the most economical choice *for certain applications* that demand high-performance lighting. If dimming functions are avoided, driver ICs can be more compact and easily integrated with the entire subsystem. In order to accommodate for low-brightness or dimmed illumination requirements, one may choose effectively to distribute several LED lamps over the desired space and simply turn off some of them when a darker ambience is preferred.

3.2.2.3.2 Dimmable LED Drivers Dimming functions are desirable to optimize illumination as ambient light levels may change due to natural or artificial lighting variations both indoors and outdoors. They are also effective in theatre and other entertainment lighting. Controlled dimming, of course, is a necessity for RGB (red, green, and blue) types of LED lamps used in electronic displays such as billboards and EMCs because pixel colors are precisely generated by means of adjusting the RGB lumen ratios. Dimming is also needed to regulate the overall display luminance based on ambient illumination levels.

Dimming of LED lamps can be accomplished primarily by either pulse width modulation (PWM) or analog techniques. In PWM dimming, the output power (i.e., the lamp lumen) is controlled by varying the duty cycle, D_{PWM}, of the input drive current, I_F, whose signal is generated using a rectangular pulse wave, as shown in Figure 3.6. The average input drive current, $I_{F(AVG)}$, is directly dependent on the duty cycle. Conventional dimmers used for incandescent lamps use TRIAC (**tri**ode for **a**lternating **c**urrent) dimming, which is also used for duty-cycle control but is applied directly to the sinusoidal AC voltage of the mains. TRIAC dimmers can be translated to LED-compatible PWM dimmers with additional circuit functions. While electronic displays exclusively require PWM, LED luminaires may use both techniques to suit different lighting applications and retrofit needs.

The analog dimming technique controls the lumen output by adjusting the drive current magnitude continuously and thus achieves full dimming, allowing output to diminish completely without creating any flicker. However, the chromatic parameter shifts are typically higher with this dimming method because the variable current continuously supplied to the LED is always a portion of the total current from the power supply. The remaining portion of the power from

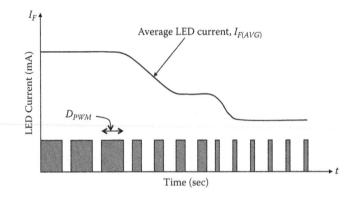

Figure 3.6. Schematic diagram of PWM dimming used in an LED lamp; the lamp is being driven with I_F, the current that flows through the LED. The average LED current, $I_{F(AVG)}$, is directly proportional to the duty cycle, D_{PWM}.

the supply always gets fed back to the generator, which absorbs it and causes heating for the luminaire.

In contrast, the PWM technique using very fast on–off pulses does not waste power usage because the LED is powered with either full current magnitude from the generator when it is on or *no* current at all when it is off. This works with LED light sources because they have very fast responses to electrical drive inputs. However, PWM can add flickering during the dimming process due to the on–off frequencies applied to the pulses undergoing phase modulation. Flickering can be largely avoided by raising this frequency well above what the human eye can detect, which is about 200 Hz [50]. A number of manufacturers offer various LED dimmer drivers to suit many applications [51–53]. Interested readers are encouraged to check for continuous industry developments from these and other driver manufacturers.

3.2.2.3.3 AC Driving of LEDs   An LED is an inherent forward-bias-only device and thus restricted to a positive DC voltage or a DC current operation. An AC input signal applied to an LED makes it dysfunctional because an AC cycle includes both positive and negative drive signal amplitudes and the diode requires a positive drive across its anode and cathode leads. Since most electric power systems provide AC outputs for mainstream applications including lighting, LED drivers must provide the AC–DC conversion. However, if two LEDs are configured back to back (i.e., antiparallel) and driven with an AC signal where one device would utilize the positive drive cycle and the other the negative drive cycle, then both LEDs would be forward biased and thus can operate normally. Such a driver need not provide AC–DC conversion but nevertheless must control the current and voltage magnitude driving the LEDs so as not to overdrive or burn them.

Controlling the input current with a simple resistor is inefficient as it causes excess heating and therefore a more complex circuit protection would be needed to operate with an AC drive. Because incorporation of such ICs is imperative, it appears that there would be little gain in using AC-driven LEDs since *both* conventional DC- and AC operations would need LED-specific drivers or electronic ICs.

The LED driver ICs, although of no small importance, are now being produced fairly inexpensively in compact sizes, and further technology improvements will invariably take place. Therefore, avoiding the usage of a common LED driver that includes a DC power supply for generating a constant current does not appear to be advantageous and may in fact add unnecessary thermal runaway challenges for AC-driven LEDs. However, Seoul Semiconductor, Inc., believes that AC-driven LEDs offer notable advantages for cost, lifetime, and footprints [54].

The many components in varieties of subsystem configurations described thus far assert that a complex optimization process is involved in designing an LED lamp or luminaire. For an LED subsystem, optimization not only is essential for building the unit but also can be important while operating the unit. When optimum control of various parameters is applied during its usage, it can maximize performance and prolong the lifetime of the LED luminaire. However, such sophisticated lighting control may be too costly for general purpose usage.

While each LED luminaire's performance can be maximized for a specific application using many subsystem options, such an approach is usually nonstandard and expensive for general lighting applications. Therefore, amid such variations of engines and subsystems, standards will need to become a necessity unless supply chains merge into a single, vertically integrated manufacturing operation and only the final product is marketed to the end user or system integrator.

3.3   Thermal Management and Lifetime Studies

In Chapter 2, Section 2.4.1, we saw that some of today's best achievements yield a total energy efficiency of 58% from blue LED chips. This alone would generate an appreciable amount of instantaneous thermal power, which is further enhanced as the applied current is passed through the diode whose resistance diminishes over time. Thus, a significant amount of heat is accumulated in the small diode active region when the LED lamp remains on for some time. Even if η_T is increased to near the practical limit described in Equation (2.19) in Section 2.4.3.2, if the lamp's usage cycle is long, an appreciable amount of accumulated heat will still need to be removed from the LED chip. Heat removal from LED lamps is crucial because, as previously discussed, light emitted from semiconductor materials has properties that depend strongly on temperature; further, thermal aging of these materials affects their light properties quite significantly as use continues.

In this section, several LED thermal issues are presented in more detail, including modes of heat transfer, the role of LED junction temperature, and reduction of thermal resistance to increase heat removal via conduction. An iterative simulation process using some measured parameters is also described to model an LED luminaire's thermal behavior and to approximate the junction temperature. Finally, some aging studies are discussed to develop a better understanding of LED lifetimes.

3.3.1   Heat Transfer Mechanisms

There are three basic mechanisms for mass heat transfer when temperature gradients are present: (1) radiation, (2) convection, and (3) conduction. Using the radiation process, an object can self-dissipate heat from its surface uniformly in all directions if the surrounding environment is cooler and homogeneous. The tungsten wire in an incandescent lamp dissipates heat primarily through radiation. In a liquid or gas, heat transfer can occur through applied convection amid a temperature gradient. Forced air or water flow, for example, can provide cooling or heating via convection for a desired object. And, finally, an object can dissipate heat by means of conduction when its surface comes in contact with another surface of different temperature.

Conventional luminaires that use incandescent and fluorescent lamps utilize a combination of radiation and convection to dissipate heat from the source. However, the most efficient method of heat dissipation for an LED chip is through conduction; heat dissipation through the radiation process in the direction of emitted light is virtually nil because the LED luminescence is concentrated primarily over a narrow wavelength range in the optical frequency spectrum and contains no thermal radiation. Thus, heat is most effectively first removed from the back of the LED chip via conduction by making a surface contact to a heat sink; then the luminaire can further dissipate heat through additional conduction and convection means.

3.3.2   LED Junction Temperature

In Section 3.2, various subsystem technologies to address thermal management were discussed. These help remove the heat from the hottest region of the entire luminaire, which is the p–n junction region—or the active layer of the LED chip, as shown in Figure 3.7. Its temperature is denoted as the junction temperature, T_J.

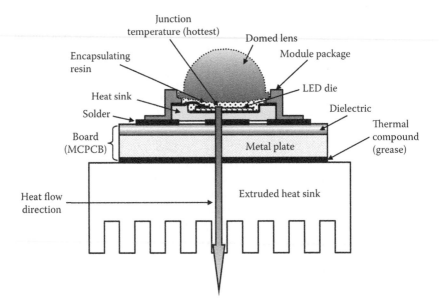

Figure 3.7. Schematic cross-section drawing (not to scale) of an LED luminaire subsystem showing the LED chip inside the module, and many other surrounding elements. The hottest point here is the LED junction inside the chip, from which the heat must be removed via conduction along the direction shown by the graded arrow.

An increase in injection current, I_F, leads to an increase in T_J and, for this reason, HB-LEDs operating with higher current require more sophisticated thermal management than 5 mm-T1-3/4 types of LEDs do.

Every LED module, due to its particular design, material properties, and fabrication qualities, has an inherent maximum T_J, which we may denote as T_{JMAX}. Overdriving this figure would lead to a catastrophic failure due to thermal runaway because, as the drive current is increased beyond a certain level, the LED's lumen output and T_J would increase greatly while its electrical series resistance would drop sharply—soon leading to a diode burnout with an electrical short. Typical recommendations are that the operating I_F should correspond to a T_J that remains 20% to 50% below T_{JMAX}, depending on the luminaire's thermal management effectiveness and usage. Additionally, similar maximum operating temperature limits should duly apply to other parts of the module or luminaire—for example, the case or the external heat sink. In sum, LED manufacturers should specify T_{JMAX} for the chip, maximum temperature for some other suitable external component of the module or luminaire, and the safe operating ambient temperatures for the lamp.

3.3.3   Thermal Analysis and Modeling

In order to design and manufacture an LED subsystem for a luminaire that can effectively remove heat from the junction region, one must first adopt a proper thermal analysis of the problem at hand. Such an analysis may be applied toward modeling and simulation of various options that must be optimized and verified by measurements to determine the subsystem design parameters and technology choices. The simple thermal analysis shown in Figure 3.8 is relevant to LED module and luminaire design.

In the simple model depicted in Figure 3.8, *material 1* and *material 2* are analogous to the LED chip and the chip submount (i.e., the first heat sink) respectively. One may add *material 3* for the PCB–metal combination, which would be the second heat sink. The junction temperature T_J is analogous to T_1, and T_2 is analogous to the temperature of the first heat sink; similarly, one may add T_3 for the PCB–metal, the bottom of which is usually the ambient temperature when a stand-alone LED module is being tested.

3.3.3.1   Thermal Resistance

Thermal resistance is defined by the resistance of a material to heat conduction, which depends on the material's thermal conductivity, *k,* as well as its length and cross-sectional area. It also equals the temperature difference between the two adjacent surfaces, where heat transfers from one to the other along a perpendicular path, divided by the heat energy flow rate, as shown in Figure 3.8. Thus, *R* is defined by

$$R = \frac{T_1 - T_2}{\dot{Q}},\tag{3.1}$$

where $\dot{Q} = \partial Q/\partial t$ is the heat flow rate of the heat energy, Q, that is present where the hotter temperature, T_1, is defined. Both definitions are analogous to electrical resistance.

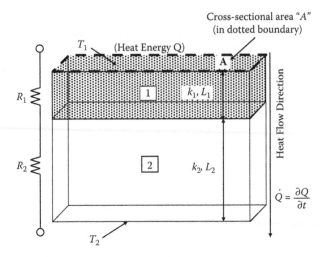

Figure 3.8. A simple thermal-analysis schematic diagram analogous to an LED chip mounted on a heat sink. Heat transfer across a composite slab of two materials, 1 and 2, experiences a series thermal resistance $R = R_1 + R_2$. Temperatures T_1 and T_2 are at the outer surface of material 1 and 2, respectively, where $T_1 > T_2$. Heat flows from the top of the composite slab where the temperature is T_1, through material 1 and then 2, experiencing a thermal resistance R.

In Figure 3.8,

$$R_1 = \frac{L_1}{k_1 A} \text{ and } R_2 = \frac{L_2}{k_2 A} \tag{3.2}$$

where k_1 and k_2 are the conductivities of slab materials 1 and 2, respectively, and L_1 and L_2 are the lengths of slab materials 1 and 2 respectively. The cross-sectional area A is the same for both slabs in this example. Note that R_1, R_2 and therefore R can be calculated if the material properties k_1 and k_2 are known along with the geometric parameters L_1, L_2, and A.

The unit of thermal resistance as defined by Equation (3.1) is degree Celsius per watt (°C/W), where "W" is thermal wattage or power. In the case of an LED, two sources contribute to thermal power that are additive and accumulative over time. One source of heat generation is when the LED is turned on with some injected electrical power (i.e., voltage times current)—only some of which is converted to emitted optical power based on the chip luminous efficiency; the remainder gets translated to thermal power. The other source that continuously accumulates increasing thermal power to the LED chip is due to itself being a resistive load in the electrical circuit. The LED thermal power should not be mistaken for electrical power when one attempts to calculate the thermal resistance in an LED module using Equation (3.1). Also, note that in a more efficient LED, the generated thermal power is less than that from a less efficient LED, even when the same amount of electrical current is injected into both.

3.3.4 Thermal Simulation

The simple analysis case in Figure 3.8 may serve as the basis for determining a suitable subsystem design that can remove the necessary heat from the LED junction area. The actual analytical problem is rather complicated because it is difficult to measure or calculate T_J accurately. However, one can perform several approximate measurements of T_J at different ambient temperatures and incorporate the results iteratively into a computer model that can then simulate the thermal behavior of an LED module or subsystem rigorously. Such a simulation process will yield good design solutions for effective heat management.

Accurate thermal simulation of an LED module or subsystem entails the application of a full three-dimensional (3-D) finite element or finite difference time-variant method to model its thermal behavior. At a basic level, it requires solving the heat equation [55] of the ensemble that includes the LED being driven with a constant current upon reaching steady state after a certain time. Each different component may be uniquely represented by its material property k, geometry, and boundary values and conditions with respect to other adjacent components and the environment. A series of calculations should be carried out for various ambient temperatures in conjunction with the corresponding measured parameters that can be applied back to the simulation to achieve convergence of the iterative process. Caution should be taken toward calculating thermal resistances appropriately for different ambient temperature cases because, as defined in Equation (3.2), k and the geometric parameter constituents are all temperature dependent.

3.3.4.1 Simulation Techniques

A number of software packages perform thermal simulations for various real-life problems. Those that simulate a wide variety of electronic components and systems, including circuit boards, semiconductor devices, heat sinks, and enclosures for complex duty cycle transient scenarios, are most suitable for LED thermal simulations when some customizations are implemented. The *Sauna*™ thermal modeling software from Thermal Solutions, Inc. (Ann Arbor, Michigan) is a computation tool that is fast and efficient in generating and solving complex 3-D models for high-power LED lamp and luminaire configurations [56].

It essentially performs finite difference *vectorial* calculations, with an object-based meshing approach of creating the necessary geometry for the problem at hand. *Sauna*™ can fully take into account all thermal properties of components and their surroundings and it automatically calculates heat transfer coefficients including convection and radiation. A user-controlled variable and optimized meshing depending on the density of high-power semiconductor components leads to a fast and accurate convergent solution. Such unique characteristics of *Sauna*™ allow one to perform essential thermal modeling and simulation of high-power LED lamps and luminaires without utilizing full 3-D computational fluid dynamics (CFD) simulation packages that consume a great deal of time and effort.

Here we present the simulation of two cases using *Sauna*™ (version 4.15), observing the effect of increasing the number and density of packaged LED modules in a subsystem. Prior to comparing these two specific cases, a good amount

Table 3.1. Vertical Dimensions of LED
Module Assembly for *Sauna*™ Simulation

Vertical Dimensions of LED Module Assembly		
Layer Description	Millimeters	
LED chip	0.1000	
Metallization for solder	0.0500	
Ceramic (99% alumina)	0.3000	
Metallization for solder	0.0500	
Copper	0.0500	MCPCB
Dielectric	0.1000	
Aluminum alloy	1.5000	

of optimization was first performed to achieve a low T_J of only 62.17°C for a single LED module on a 30 mm × 30 mm MCPCB mounted on a finned heat sink. The fin spacing and length are 5.5 and 30 mm respectively. The size of the LED chip is 1 mm × 1 mm, which is soldered onto 4 mm × 4 mm alumina ceramic. The input thermal wattage of a single LED chip is taken to be 1.3 W. The vertical dimensions of the ceramic and PCB are provided in Table 3.1.

Figure 3.9 shows the thermal behavior of a four-module assembly on the MCPCB with a center–center spacing of 13.0 mm. The increase in T_J from one

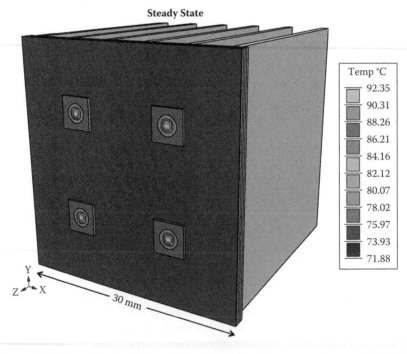

Figure 3.9. *Sauna*™ simulation of four LED modules mounted on an MCPCB. The calculated T_J is 92.35°C.

Steady State

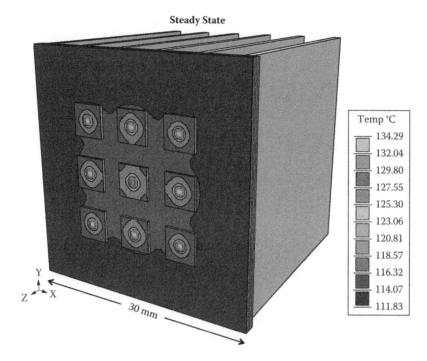

Figure 3.10. *Sauna*™ simulation of nine LED modules mounted on an MCPCB. The calculated T_J for the center module is 134.29°C, whereas the T_J for the peripheral modules is 133.30°C.

module to 4 modules is 30.18°C. Figure 3.10 shows the simulation for a nine-module assembly with a center–center spacing of 6.5 mm. The increase in T_J from the four-module case is 41.94°C! In both cases, however, T_J remains well under T_{JMAX} (typically higher than 150°C) for most well-designed high-power LEDs despite such close proximity of the modules. Both cases show symmetrical behavior in the thermal contours following the symmetry of the assemblies. In Figure 3.10, the center module shows a slightly higher T_J than the other peripheral modules do. The temperature difference is only 0.99°C. The reader is encouraged to seek why the center module has a slightly higher T_J.

3.3.4.2   Incorporation of Thermal Measurements

Thermal analysis and design depend critically on the accurate knowledge and determination of the quantity T_J belonging to an LED device. Unfortunately, the LED junction is not accessible after encapsulation and therefore direct measurement is unattainable. It is best for LED chip manufacturers to provide a reasonably accurate value of T_J by means of some covalidation between the diode behavior calculation and its chip-level measurement. Although it is unrealistic to do that for each LED device, with improved manufacturing yield and reduced binning variation, an average rating for T_J from chip and module manufacturer should be helpful for a particular series of LED engines in the future. Currently, the better choice may be to obtain a functional dependence of forward voltage (V_F) versus temperature from the LED manufacturer and then to use an extrapolation method to estimate T_J by measuring V_F at various temperatures, after a thermal equilibrium has been reached. The simulated model described previously can also be utilized to estimate T_J by making analytical comparisons between the measured and simulated temperature rises of a suitable point on the PCB or heat sink. Both methods rely on some complex functional dependence of T_J on the LED's material parameters, such as doping and defect densities and other device parameters that determine its luminous efficiency.

The science and engineering behind determining T_J and T_{JMAX} with good accuracy are quite complex and time consuming. It requires an iterative effort between rigorous modeling and extensive measurements of various LED parameters at different ambient temperatures. It may require many lengthy iterations to ensure the convergence of a valid and convincing T_J value. Quantifying relationships between T_J, T_{JMAX}, and optimal drive current for various LED modules to be used in different applications are currently important research and development areas for the SSL industry.

3.3.5   Lifetime and Aging Studies

Determination and assessment of an LED lamp lifetime are unique and complex in comparison to those for incandescent and fluorescent lamps. Nonetheless, by and large, the LED industry, as well as casual observers, currently believes that LED lamps last for a *very* long time. The "LED lamps operate interminably, or almost forever" belief has unfolded because many LED manufacturers claim a 100,000-hours life span. Although such claims and beliefs have become popular, many contrasting cases tell a different story. A classic example is the recently upgraded, partially darkened LED-arrayed traffic lights often seen on roads today. If ordinary traffic signal LEDs experience life span inconsistencies, the SSL industry has an obligation to clarify expectations of the not so well tested, especially the much lauded, white HB-LEDs.

3.3.5.1   Defining LED Lamp Lifetime

In the lighting industry, when a lamp's lifetime is specified to be 3 years, the professionals recognize that only half the lamps of that series are expected to last that long. The remainder may only last any fraction of 3 years. We may call this the "second-half" syndrome. Too often, the second-half failure syndrome frustrates many luminaire producers and end users, especially if the LED lamps burn out within months or days following installation.

There is some rationale behind why some scientific communities as well as many other people believe LED lamps last a long time or have such potential longevity. One simple reason is advertising claims; another is that LED indicator lamps on early electronic gear still produce light after several decades of heavy usage. Further, credible, laboratory tests analytically utilize the accelerated life span results of current LED lamps to determine extrapolated lifetimes that exceed 100,000 hours.

While the public generally views such published laboratory results based on extrapolation as a universal phenomenon, lighting professionals' views differ from that of the public. They perceive, for example, that a manufacturer's production run of two million, 100,000-hour lifetime LED lamps produces roughly one million that will endure. We may describe the long-life lamps as "first-half lamps," because they prevail.

LED lamp lifetimes cannot be generalized for all series that support a wide variety of applications. For example, the LED indicator lights in electronic or electrical gadgets have been shown to last over several decades, which exceeds well beyond 100,000 hours. However, indicator lamps consume only a few milliamps to operate and there is little concern if the light output diminishes over time. High-power and high-brightness LEDs require high current and high current densities to operate, which increases the diode-junction temperature, leading to faster aging and a shorter lifetime. In general-illumination and other high-brightness applications, single LED emitters are required to produce several hundreds of lumens from limited size sources. Additionally, these types of lamps demand certain luminance levels as well as color quality and uniformity. Certain high-end applications, such as automotive headlights, require even higher luminance and lumen outputs, which invariably demand very sophisticated, thermal management systems to ensure longevity with high optical quality. For example, Osram Sylvania produced its OSTAR brand LED headlights for Audi's $75,000 A8 sedan. The company specified a minimum life of 7,000 hours for the headlights [57]. While this life span deems adequate for this application, LEDs with much lower luminance and flux typically have much longer lifetimes.

For general purpose LED lamps, specified lifetime ratings are meaningful if manufacturers test a production run of a large number of similar LEDs for all relevant light characteristics and demonstrate that half, at least, endured as expected. Explicitly, no light characteristics would have dropped below published specifications. Currently, the lighting designers find that this is rarely the case.

To justify advertised lifetimes, diligent manufacturers perform certain equivalent tests via some sampling, as well as accelerated temperature and aging tests. Due to the low yield still being prevalent for high-performance HB-LEDs, such claims may be valid for smaller quantities—typically for clusters of 100, rather than for tens of thousands or millions. This is because equivalency tests and samplings do not always scale, since significant, temperature-related, performance variations exist even among LED devices built from the *same* compound semiconductor epitaxial wafer, as well those from like wafer batches. Other variations such as the thermal management quality will also cause lifetime deviations within the entire lamp manufacturing line.

3.3.5.2 Lumen Maintenance during Lifetime

Most manufacturers of high-power, white LED devices and lamps now estimate a lifetime between 30,000 and 50,000 hours, with an expectation that at least 70% of the initial lumen level will be maintained for this life span. This estimate generally assumes operation near 350 milliamps (mA) of constant current and a junction temperature not exceeding 90°C. Some research results have produced longer-life LEDs that, to some extent, tolerate higher drive currents and operating temperatures. Presently, certain manufacturers offer LEDs rated for 100,000 hours while operating at over 700 mA with junction temperature up to 120°C [58].

The SSL industry primarily considers 70% and 50% lumen maintenance levels as life span definitions, but for different applications [59]. These lifetime specifications are inevitable and meaningful because an LED lamp's efficiency drops over time due to gradual heating, causing the flux output to diminish slowly. The Illuminating Engineering Society of North America (IES) test standard LM-80 is the approved method for measuring lumen depreciation for LED light sources, arrays, and modules; however, it does not cover measurement of luminaires, which of course includes drivers and other components [60]. The IES technical memorandum, TM-21, specifies the extrapolation of LM-80 data to make lumen maintenance projections in time going beyond measurement data [61].

Although LM-80 and TM-21 are not standards for defining lifetime, many manufacturers declare lamp expiration based on when the lumen output falls below 70% (or 50%) of the original output. Interestingly, the average person is unable to judge when the LED lamp has reached this point, which is the drawback of this custom. In contrast, an incandescent light bulb ends its usefulness when the filament breaks and the light ceases.

Luminaire designers need to better understand LED lifetimes when they design LED subsystems and thus need to ask LED emitter or engine manufacturers for life span data sheets that describe the gradual behavior of the luminous-output degradation over time and under specific operating conditions. Specifically, data relating to T_J and T_{JMAX} are crucial as they are used to determine specific LED-based luminaire design and lifetime. Finally, lamp and luminaire specifications should include maximum ambient temperature ratings. High ambient temperatures and humidity can potentially lead to a reduction in life span.

Determining valid LED lamp and luminaire lifetimes is a complicated process and the standard bodies are working to release an update to LM-80 that promises to provide considerable value to the industry. Because LED-based lighting product lifetime is based on projections from certain accelerated aging tests, it is important that the testing conditions used in such tests are appropriate in order to be meaningful for the actual product applications. A great deal of scientific and technological methods relating to lifetime studies on solid-state light sources and systems from the prior telecommunication industry standards leader Bellcore (formerly part of Bell Technical Laboratories of AT&T and now Telcordia), for example, may be adopted toward LED lighting products. However, suitable translations with respect to lighting applications must still be incorporated into such parameters as test cycling, duration, and other applicable environmental and mechanical conditions.

Finally, concerning LED lamp longevity, it is important to reiterate that development and implementation of effective thermal management technologies will ensure very long lifetimes and also yield more uniform products. This will substantially reduce binning requirements for LED lamps that greatly concern the customers today. Because semiconductor properties are inherently sensitive to thermal variations, it is also important that LED lamps are carefully treated with respect to environmental thermal issues during their lifetime so that stable LED light characteristics can be achieved.

3.4 Optimizing Module Designs for Manufacturing Platforms

As discussed in Section 3.2, LED lamps and luminaires are an intricate ensemble of many parts and configurations involving thermal, optical, electrical, and mechanical aspects. Their interdependencies simultaneously affect the lamp's efficacy, brightness, color quality, stability, and longevity. In order to achieve desirable and reliable performance for these lamps, it is important to understand how acutely the lighting characteristics vary with chip design dimensions, material qualities, fabrication variability, and packaging methods. Therefore, to produce superior quality lamps, an LED engineer will need to optimize lamp designs using rigorous thermal, optical, electrical, and mechanical engineering procedures, while ensuring harmonization with manufacturing requirements. We now address the design aspects for these four disciplines.

3.4.1 Thermal Design Considerations

The SSL industry has taken on the thermal challenges seriously as individual LEDs are becoming brighter, bigger, and able to consume more electrical power before experiencing a burnout. LED technologies continue to improve with increasing demands for HB-LEDs for many new and existing lighting applications. In order to meet the demands of various applications, LED module and luminaire developers must consider certain thermal design methods that can substantially boost their product performance.

In order to conquer the core of thermal problems, LED engineers must resolve how to reduce the dependence of T_J on the operating (drive) current at the onset, which, in turn, will lead to increasing LED lamp life span. As previously mentioned, a more efficient LED chip will produce less heat in the junction region than a less efficient LED would when both are driven with the same amount of injected current. Stronger T_J dependence on drive current may also lead to faster and greater degradation of color quality, due to wavelength shifting, which becomes more pronounced over time. No doubt, future design improvements in semiconductor epitaxial quality, LED device design, metallization, and other related processes will lead to substantially lower T_J dependence on operating current at the onset. Nevertheless, a lower operating current can always significantly increase LED lamps' lifetime; lower currents also simplify thermal management and reduce design and manufacturing costs.

With a sophisticated thermal management scheme incorporating both structural designs and packaging and material technology, higher T_J at the onset can be tolerated. It can be achieved by creating an efficient thermal conduction path from the LED junction area to the immediate heat sink on which the LED is mounted, then to the printed circuit board (PCB) and the second heat sink, and finally to the external environment for the luminaire. This path is shown with a downward arrow in Figure 3.7. This will allow T_J not to rise excessively as the lamp stays on and a thermal equilibrium can be reached where T_J stays well below T_{JMAX}. Creating such a low thermal series resistance along this conduction path constitutes the usage of appropriate thermally conductive materials (e.g., high thermally conductive substrates such as alumina and metal core PCBs and good thermal epoxy or grease), packaging techniques (e.g., good solder joints), and substrate geometry that may include thermal via holes and flat mating surfaces. These elements invariably increase the cost of the modules and luminaires and most often increase their sizes as well—both adding to concerns for general illumination uses. These can lessen if the individual LED modules themselves generate less heat.

Some practical means that help create an efficient thermal conduction path from the LED chip to the external heat sink of the luminaire are described next.

3.4.1.1 Module Heat Sink

An immediate heat sink to the LED chip is crucial to its heat management. It is the first element that connects the bottom of the LED device surface to a bigger and cooler surface area to dissipate heat gradually and effectively to the external environment. Although one can benefit from such active cooling methods as forced convection using airflow from a fan or cooled water flow or a thermoelectric cooler, a passive heat sink is much simpler and less costly. It can be ceramic or metal depending on applications. The effectiveness of the heat sink improves with larger and flatter surface area as well as secure connections to PCBs.

3.4.1.2 Board Technologies

The PCB technologies were discussed in Section 3.2.2.1. Here, we highlight the important points once more. Most high-performance HB-LEDs use MCPCBs, which are also called "insulated metal substrates" (IMSs). LED emitters or modules are soldered directly to metal cores of MCPCBs to maximize heat transfer from the LED heat sink to the PCB. To ensure that heat transfer is effective for the entire luminaire, designers must determine an adequate PCB size, as well as spacing of active components, including LEDs and driver ICs, because they produce heat during operation. Depending on the application, some small-wattage LED lamps' thermal requirements may be met using only an embedded heat sink in the module and an appropriate PCB design. However, for large wattage LED lamps and luminaires, the need for enhanced thermal management with ceramic submounts for individual LED modules and extruded heat sinks for the PCB become imperative.

3.4.1.3 Extruded Heat Sink

If lamps or luminaires are required to produce very high luminance and flux in high-temperature environments, elaborate heat sinks are usually utilized to remove heat adequately away from the LEDs. The most common technique is to use a heavyweight aluminum extruded heat sink that provides significant surface area for dissipating the heat into the environment that originates

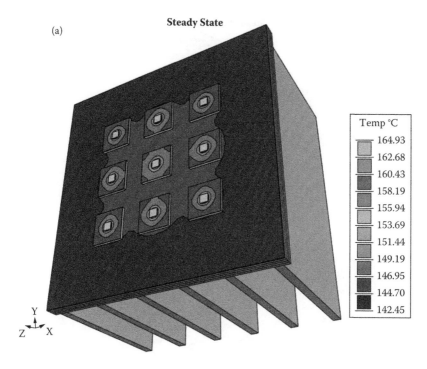

Figure 3.11. A simulation study illustrating the improvement in thermal management using longer cooling fin lengths in an HB-LED-based lighting subsystem that uses nine 1 mm × 1 mm LED emitters mounted on an MCPCB, spaced by 5.5 mm: (a) with 18 mm fin length for the heat sink; (b) with 30 mm fin length—showing a T_j reduction over 30°C from the case in (a)! (The simulations were performed using Sauna Thermal Analysis Software from Thermal Solutions, Inc.)

from the LEDs and then transferring to the heat sink fins via the MCPCB. In Figure 3.11, thermal simulations of two cases of HB-LED arrays, each consisting nine modules mounted on an MCPCB are shown, differing only in the fin length of the extruded heat sink below the MCPCB plane. (Detailed parameter descriptions of the LED assembly, thermal power, and others are given in Section 3.3.4.1.)

The simulation results presented in Figure 3.11 show the effectiveness of longer fins. Practical demonstration of such design improvements is seen between the 60 and 75 W LED A-line retail lamps by Philips, where the latter uses an extruded heat sink with nearly 30% longer fins.

If the lamp or luminaire spatial conditions allow it, a suitable aerodynamic design may be applied to the heat sink in order to improve air circulation. Many passive metal heat sinks of this kind are constructed using such tooling techniques as extrusion, casting, milling, stamping, and bending. Figure 3.12 shows a curved, aerodynamic extruded heat sink in an 8 W LED light bulb from Osram Sylvania, which has an L70 lifetime rating of 50,000 hours.

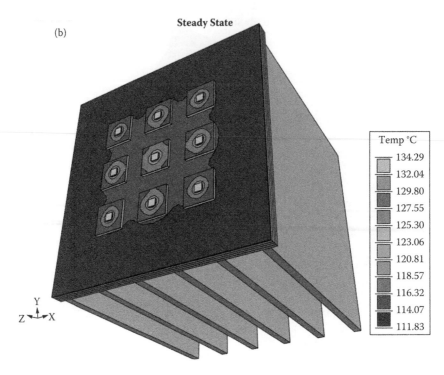

Figure 3.11. (continued) A simulation study illustrating the improvement in thermal management using longer cooling fin lengths in an HB-LED-based lighting subsystem that uses nine 1 mm × 1 mm LED emitters mounted on an MCPCB, spaced by 5.5 mm: (a) with 18 mm fin length for the heat sink; (b) with 30 mm fin length—showing a T_j reduction over 30°C from the case in (a)! (The simulations were performed using Sauna Thermal Analysis Software from Thermal Solutions, Inc.)

3.4.1.4   Adhesive Materials

The thermal interface material and adhesion technique used to bond PCBs and heat sinks play an important role in maximizing thermal conductivity in an effort to reduce the LED junction temperature. Although PCBs and heat sinks are designed to have homogeneous and flat surfaces, the manufactured versions typically have surface imperfections that add to thermal resistance due to small air holes. Such effects can be minimized by using high-quality thermal bonding adhesives to mate the surfaces evenly by eliminating pockets of air along the interfaces. Strong adhesion bonds are also necessary to mount the large and heavy heat sink securely to the PCB.

3.4.1.5   Utilizing Convection

While thermal conduction is the primary mode of transferring heat in an LED lamp, convection can also help in several ways. The thermal power that is removed

Figure 3.12. An 8 W, 120 V E26 LED light bulb from Osram Sylvania showing external, extruded metallic heat sink under the emitter. This bulb has a lifetime rating of 50,000 hours for 70% lumen maintenance. The retail cost of this bulb was $46.33 in low quantities in May 2012. (Photo courtesy of Osram Sylvania.)

from the LED chip to the external heat sink dissipates heat into the environment through convection, which may be enhanced by utilizing an aerodynamic heat sink design as discussed at the end of Section 3.4.1.3. In addition, when possible, a large number of vent holes should be machined into the luminaire housing to create effective air flow to dissipate heat through convection into the surrounding environment. A small amount of heat also dissipates through radiation in LED luminaires. These concepts are shown in Figure 3.13.

3.4.1.6 *Optimization of LED and Drive Current Quantities*

The selection of LED emitter quantity and nominal drive current for a luminaire is vitally important for achieving high-quality thermal as well as optical performance. When higher and more distributed illumination is preferred over a large space, it is better to increase the number of individual LEDs within a luminaire and, simultaneously, to reduce the drive current to generate less heat [62]. Many lighting applications may be addressed with this type of optimization, which should take some of the burden off the thermal management scheme and offer better light quality and longer life span.

To illustrate this principle, it should be possible to construct an LED luminaire with only relatively simple passive cooling technologies that may replace a 13 W CFL by using 10 emitters, each operating at 350 mA, to produce at least 100 lm. At a typical operating temperature of 25°C, the junction temperature could be kept at or below 70°C for the 10 LED emitters, which is far lower than the usual T_{JMAX} ratings of 150°C from certain prominent HB-LED manufacturers [63]. Such

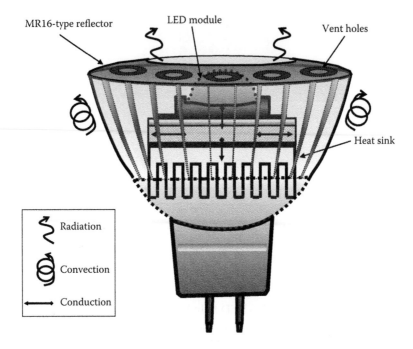

Figure 3.13. Schematic illustration of an MR16-type high-power LED luminaire design showing vent holes in its housing for improved thermal management. The LED engine inside can be seen as the luminaire housing is made transparent for illustrative purposes. All three types of heat transfer mechanisms for the luminaire are shown in the illustration.

a conservative thermal design for similar LED luminaires may provide the ability to produce over 1000 lm, while ensuring the long life expectancy and minimizing color shift by avoiding operation at high junction temperatures.

3.4.1.7 Active Cooling

Thus far, only passive cooling technologies have been discussed; these are generally the preferred methods for LED fixtures because they do not require any energy or moving parts for their operation. While passive cooling is more practical and cost effective for most regular usage in stand-alone luminaires and lamps and for small lighting systems, active cooling may be necessary for large systems that require sophisticated color and intensity controls. This is usually observed in entertainment industries—for example, in sports arenas, building facades, and theatres. The newly constructed Dubai stadium in Abu Dhabi is a good example that is known to utilize occasionally active cooling for a gigantic number of LED emitters that help provide grand entertainment lighting in a warm climate region. The active cooling techniques are also used for line scanning in machine vision where an extremely high level of light is demanded; Schott (North America) has used active air- or water-cooled white and red LEDs to produce up to 2 million lux (lumens per square meter) in a line scanner [64].

LED lighting systems, employing active cooling and various intelligent controls in interpretive arrangement, degrade at a much slower rate. However, to come to common use, LED lamps for regular household lighting and other applications need less costly performance and long lifetimes, without sophisticated systems or end-user adjustment.

3.4.2 Optical Design Considerations

Optical performance is the most defining and recognizable feature of a lamp. When speaking in abstract terms, it may seem to the end user that a light bulb is a light bulb and an illuminated room is what it is. However, a light bulb's real ability to illuminate more effectively and a well-illuminated room carefully designed by a lighting professional are both distinguished, appreciated, and recognized by many of us. Prior discussions leading up to now provide an understanding of some intricacies of multidisciplinary science and technologies behind producing light radiation from LED lamps. Since optical design elements are a major part of the following chapters of this book, we shall briefly outline the overall design considerations in this section.

3.4.2.1 Light Extraction from LED Emitter

Maximizing light extraction from the LED chip requires minimizing optical scattering and absorption inside it. It entails creating smooth epilayer interfaces and low defect densities in the semiconductor materials to decrease scattering and absorption. The roughness in metal and insulator interfaces within the fabricated chip also contributes to optical loss and therefore must be minimized. However, the smoothness on the top-most epilayer can create substantial reflection based on the large index mismatch between it and its surrounding, which may be encapsulating resin, phosphor layer, or air. This reflection causes the generated light to reenter the chip and get absorbed within. Hence, this surface reflection must be reduced by employing index matching techniques such as incorporation of quarter-wave stacked layers or some corrugated geometrical structures that create periodic surface roughness.

Similarly, phosphor loss must be minimized by increasing its homogeneity and efficiency. Depending on where the phosphor is placed within the light engine, light extraction from the entire ensemble must be maximized by reducing unnecessary reflection, absorption, and scattering from all secondary optical elements.

3.4.2.2 Shaping Light Distribution

Spatial illumination characteristics play a major role in a wide variety of lighting applications including general purpose everyday lighting. This is previously discussed as flux distribution, $\Phi(x,y,z)$. LEDs manufactured with current methods generate a near-field distribution that can be approximated with a Lambertian and thus are mostly suitable for illuminating flat surfaces in *very* close proximity. Although casually described as directional light sources, they are still only useful for such short-distance applications as LCD backlighting and task lighting. Their directionality is not useful for large distances like that of a laser beam or even for a flashlight type application; LED emitters must use a beam collimator to provide useful directional illumination for medium distances.

The desired spatial distribution can be generated in a variety of ways for an LED luminaire. Several of these secondary optics technologies will be discussed in Chapter 6. Effective secondary optics can generate the desired light distributions for specific applications.

3.4.2.3   Color Parameters

The complete optical design consideration of an LED lamp or luminaire includes meeting the requirements for color rendering index (CRI) and correlated color temperature (CCT) color parameters, along with achieving high chip extraction efficiency and the appropriate spatial light distribution. For many applications, most of these parameters need to be traded off against one another and therefore optimization is necessary to achieve the desired performances from a particular LED luminaire.

3.4.3   Electrical Design Considerations

Electrical driver utilization has been described in Section 3.2.2.3. Here we consider how an LED lamp or luminaire's performance can be enhanced by optimizing several electrical parameters.

3.4.3.1   Fabrication of Anode and Cathode Contacts

As we have seen with optical and thermal designs, the main electrical performance is also related to the optoelectronic properties of the LED chip. The chip's I–V (current–voltage) and L–I (light–current) characteristics improve with high-quality fabrication techniques that invariably produce robust and stable electrical contacts with low diode series resistance. The total series resistance should be minimized using highly conductive metal alloys to form short wire bonds and external leads.

3.4.3.2   LED Arrangement Designs

The electrical driver design depends on how many LED emitters are used in a particular arrangement to suit a certain lighting application. For example, if a lamp used a significantly different arrangement of varying numbers of LEDs in series or parallel connections from another to produce different flux distribution and luminance, its driver efficiency and power factor would not be the same as the other's.

LED lamps' endurance is directly connected to electrical driving conditions. Because low- and high-power LEDs have substantially different I–V and L–I characteristics, suitable operating currents must be chosen to avoid premature burnouts. Expectedly, all diodes burn out if one applies excessive current. In order to optimize the driver efficiency as well as to avoid thermal runaway, an LED emitter arrangement needs to be properly wired. Properly choosing voltages, resistors, capacitors, and currents within a driver architecture can allow both series and parallel connections and combinations. If LEDs are connected in parallel, *all* emitters must be sufficiently identical with respect to the operating current, resistance, and diode turn-on forward voltage, V_F. However, this requires proper binning of equivalent LEDs. Alternatively, an appropriate electrical driver may provide compensation for different V_Fs. Red, blue, and white LEDs inherently require different voltages due to different energy gaps, as discussed in Chapter 2. Therefore, parallel connections will require each color LED to have its own series resistance, to limit the current passing through the diode.

3.4.3.3   Electrical Control Functions

A number of features of an LED luminaire can be controlled with its electrical driver. The most notable one is the dimming of the lumen output as discussed in Section 3.2.2.3. The addition of a dimming feature may affect the PFC adversely. So it is important to make sure that while dimming is applied, PFC is not traded off substantially. Otherwise, it would defeat the purpose of trying to save energy on one hand via dimming the light, while using up much of the line power to operate a dimmer.

An example of an electrical driver that controls various features of a certain category LED lamp is National Semiconductor Corporation's (now Texas Instruments) LM3433. It is a common-anode, current-mode driver used in high-power, high-brightness LED backlighting, miniaturized projector, and lighting applications. The LM3433 is a DC–DC buck-based constant-current regulator. Its short, constant-on-time architecture uses small external passive components without an output capacitor. It uses both analog and digital current control modes for adjusting manufacturing variations and PWM dimming respectively. Other features include thermal shutdown, VCC under-voltage lockout, and logic-level shutdown mode [65].

Electrical drivers also offer numerous control functions for such full-color displays as back-lit LCD televisions and LED-based billboards and EMCs. These include brightness adjustments, gamma correction, color gamut extension, and optimum thermal management.

3.4.4   Mechanical Design Considerations

Mechanical characteristics are vital to an LED luminaire's longevity and handling. At the chip level, they are micromechanical aspects that include adhesion of metallization bond pads, wire bonding strength, quality of resin encapsulation deposition techniques and materials, and others. At the subsystem and luminaire level, they are macromechanical aspects, which include integration of LED engines with PCBs, drivers, heat sink structures, and others. Mechanical design and engineering of these various elements involve determination of appropriate geometry and hardware construction methods and implementation, to help minimize mechanical stress during assembly, installation, and operation. Product development and manufacturing procedures must also incorporate comprehensive mechanical tests involving shock, vibration, and liquid emersion to qualify final products that can be featured with certain lifetime warranties. Such technology development will minimize product degradation over time, offer ease of use, and protection from environmental damage. Technologies of this kind have been developed for other optoelectronic industries such as lasers, detectors, and transmitter and receiver subsystems used in communication industries [66].

LED lamp and luminaire engineers would offer great benefits to the SSL industry if they optimized designs simultaneously, covering all four engineering disciplines: thermal, optical, electrical, and mechanical. The goal of this type of design optimization should include quantifying performances and determining tolerances or error margins for all LED lamp and luminaire parameters that can be translated to manufacturing lines' abilities to generate certain levels of accuracy. The developers should use iterative processes to verify the matching capabilities of design and manufacturing techniques. They should demonstrate that a successful final product's variations are within the designed tolerances. The technology boundaries may then be extended by adopting the next goal to generate further optimized designs that will lead to higher tolerances for manufacturing products. Such an achievement would invariably lower the cost of LED lamps and luminaires.

4 Lamp Measurement and Characterization

4.1   Introduction

Despite being an intricately unique lighting technology, the LED lighting industry has made notable progress with respect to characterization of light properties in the last several years. However, due to a wide variety of configurations and applications of LED lamps and luminaires, their characterizations still lack completeness, standards, and specificity. In order to determine whether LED lamps are suitable for existing lighting applications, they must be carefully characterized to make meaningful comparisons with incumbent products.

Such diligent evaluation requires that LED lamps be first *fully* and accurately characterized for *all* the general lighting parameters discussed in Chapter 1. To comprehend such performance characterizations against incumbent lamps as well as those required for new applications, we shall investigate in this chapter how standard photometry and colorimetry may be applied to LED lamps. Then we shall analyze how they may be compared with corresponding features of incandescent and fluorescent lamps. In the final section of this chapter, we shall examine the LED-specific measurements and characterization because they remain very important for ongoing improvement of the field.

4.2   Measurement and Characterization of General Lighting Parameters

In Chapter 1, we have seen that quite a few lighting parameters affect quality and effectiveness with regard to illuminating space, whether it is for creating certain ambiance or for performing specific tasks that require certain visibility. Let us revisit these parameters and investigate which of them can be measured with what level of accuracy and how such data can be applied to a few common applications.

4.2.1   Primary Lighting Metrics and Measurements

The basic lighting metrics were provided in Table 1.3 in Chapter 1, which we now expand into Table 4.1 to include measurability by means of a few common apparatuses available in the market. These lighting metrics and measurements fall in the photometry and colorimetry categories. *Photometry* is distinguished from *radiometry* as being confined only to measurements of light energy visible to the human eye, whereas radiometry encompasses measurements of all optical radiant energy including the visible, infrared, and ultraviolet spectra. The entire discipline

Table 4.1. The Basic Lighting Metrics and Measuring Apparatus

Items	Physical Quantity (Symbol)	SI Unit	Unit Symbol	Apparatus & Vendors (When Applicable)
1.	Luminous flux (Φ)	lumen	lm	Integrating sphere; GL Optic, Instrument Systems GmbH, others
2.	Luminance (L)	lumen per steradian per square meter	$lm/(sr\text{-}m^2)$	Konica Minolta LS-100/110 and CS-100/200, others
3.	Illuminance (E_v)	lumen per square meter (lux)	lm/m^2 (lx)	Konica Minolta CL-500A, GL Optic, others
4.	Exitance (M)	lumen per square meter	lm/m^2	Konica Minolta LS-100/110 and CS-100/200 (obtained from luminance)
5.	Flux distribution ($\Phi(x,y,z)$ in Cartesian or in polar coordinates)	lumen	lm	Goniometer; Techno Team Bildverarbeitung GmbH, Instrument Systems, others
6.	Spectral power distribution	watt per meter	W/m	Konica Minolta CL-500A, Gigahertz Optik, others
7.	Lamp efficacy	lumen per watt	lm/W	Not directly measured; see Section 4.2.4.6
8.	Luminaire efficiency	None	None	Not directly measured; see Section 4.2.4.6
9.	CIE color coordinates	None	None	Konica Minolta CS-100/200, Konica Minolta CL-500A, GL Optic, others
10.	Color rendering index (CRI)	None	None	Konica Minolta CL-500A, others
11.	Correlated color temperature (CCT)	kelvin	K	Konica Minolta CL-500A, others

Notes: The measurement instruments and vendors named here are but a few examples. Other instruments and vendors exist in the market and some, which measure some of the parameters in this table, may be under development. Instrument Systems is now part of Konica Minolta.

of optical measurements is primarily subdivided into photometry and radiometry. It is important to note that, while both types of measurements can be applied to common lamps, photometry is customary and most relevant. Nonetheless, it is essential to understand the distinction and conversion methods between the two, which is particularly important for LED lamps. The conversion methods for radiometric and photometric quantities are described in Section 4.2.3.

4.2.2 Secondary Lighting Parameters

While Table 4.1 provides the lighting parameters that can be directly measured or straightforwardly calculated from measured parameters, it is important to remember that several other parameters that also affect lighting quality are more difficult to measure or calculate. These are listed in Table 4.2 as secondary lighting parameters.

Although the quantities in Table 4.2 are not always measured directly or rigorously implemented in designs for improving illumination, experienced lighting designers are able to achieve high-quality performance for these parameters by utilizing prior knowledge and qualitative analysis. If one is not vigilant in designing LED lamps and luminaires, these parameters may become more complicated to define and adjust for illumination applications compared to those for incandescent and fluorescent lamps. However, LEDs usually provide more advantages in these regards for signs and electronic displays, either inherently or through technical control schemes.

4.2.3 Conversion Methods for Radiometric and Photometric Quantities

Photometric parameters are quantified by first measuring the corresponding optical or radiant power emitted from the light sources or impinging on a particular surface, and then applying a conversion to account for the average human eye sensitivity. All photometric parameters include luminous flux in units of lumen, which, as we have seen in Chapter 2 (Equations 2.10 and 2.12) is determined by weighting the spectral radiant power measured (in units of watts per meter) with the luminosity $V(\lambda)$ function. Every photometric quantity has its radiometric equivalent that is based on watts (W) (i.e., energy per unit time). International Commission on Illumination (CIE) regulations have therefore chosen the symbols for radiometric quantities to be denoted with the subscript "e" for "energy" [69]. Thus, the analog of luminous flux, Φ, is denoted as radiant power, Φ_e. Since radiometric power is a function of wavelength, its spectral behavior is denoted as $\Phi_\lambda(\lambda)$ and is known as "spectral radiant power." The quantities Φ_e and $\Phi_\lambda(\lambda)$ are related as follows:

$$\Phi_e = \int_0^\infty \Phi_\lambda(\lambda)d\lambda \qquad (4.1)$$

This is graphically shown in Figure 4.1.

Table 4.2. Secondary Lighting Parameters and Descriptions

Physical Quantity	Description
Color balance	If the source luminance or brightness changes appreciably, individual colors of the source spectrum need adjustment or "balancing" relative to one another for best illumination or display characteristics. For electronic display applications, an algorithm called "gamma correction" is used to achieve color balance.
Acuity	This describes how clearly we see illuminated objects or display images and therefore relates to focus or sharpness. Excessive brightness and lack of color balance lead to blurry vision or image distortion for illumination and display applications.
Contrast	This describes color distinction among various colors. Because the many colors we are able to see can be composed of a few dominant colors such as RGB, color distinction relates to how effectively the dominant colors are or can be combined using proportions. Although both may have similar visual effects at times, contrast *is* different from acuity. Acuity gives clarity and sharper outlines between differing objects, while contrast provides color distinction.
Glare	This relates to unwanted or excessive reflection that adversely impacts our vision. When a point light source radiates isotropically and produces illuminance that does not exceed the human eye comfort level, it usually does not produce glare unless a portion of its light is incident on a highly reflective surface at certain angles and starts to concentrate light in particular directions that reach the eye. LED lamps with high luminance, unless created to provide isotropic illumination, will likely create more glare incidences than incandescent and fluorescent lamps.
Interference	White light does not technically cause interference; if LED luminaires use several individual monochromatic or near monochromatic LEDs of certain properties, this could create some level of undesired interference within the lamp configuration, leading to nonoptimal lamp illumination output.
Aberration	Aberrations of the human eye vary among individuals and produce vision distortion that may be somewhat dependent on the illumination wavelength. Typically, these dependents are rather small [67].
Polarization	Polarized illumination and detection can be utilized to enhance the visibility of targets obscured in highly scattering media [68].
Time variance	AC LEDs or PWM dimming utilize time variant input electrical signals. Caution must be taken for certain frequencies used relative to eye sensitivities.

Notes

1. Since optical wavelengths are on the order of a micron (1 μm = 10^{-6} m), the wavelength domain typically uses nanometers (nm) or angstroms (Å) to account for sufficient resolution in smaller units. 1 nm = 10^{-9} m; 1 Å = 10^{-10} m.
2. In practice, the limits of integration in Equation (4.1) are the boundaries of the wavelength region of interest. For example, for a typical visible range, the lower and upper limits are often taken to be 380 and 780 nm respectively.

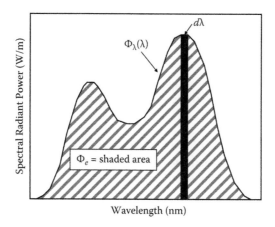

Figure 4.1. A hypothetical example showing the relation between spectral radiant power, $\Phi_\lambda(\lambda)$, and total radiant power, Φ_e.

Equation (4.1) illustrates that the total radiant power of an optical source includes all the spectral power contributions of that source. Because optical sources can be monochromatic, nearly monochromatic, and polychromatic with varying spectral behavior, their corresponding conversion between the radiometric and photometric quantities also varies.

4.2.3.1   Monochromatic Radiation

If a source emits monochromatic radiation at a particular wavelength λ in the visible spectrum, its radiometric power Φ_e (in watts) is easily converted to its corresponding luminous flux Φ (in lumens) by multiplying it with the respective $V(\lambda)$ value and the factor 683 lm/W. Thus,

$$\Phi = \Phi_e \cdot V(\lambda) \cdot 683 \tag{4.2}$$

where Φ and Φ_e are expressed in lumens and watts respectively.

Many visible LEDs in bare die form are nearly monochromatic. As long as their spectral width is not too large and does not fall in the region where $V(\lambda)$ varies significantly, Equation (4.2) provides a good approximation for converting their radiant power to lumens. It is important to note that while red and blue LED emitters may have the same energy efficiency, they will have two significantly different luminous efficacies because the $V(\lambda)$ values for red and blue shades are very different.

4.2.3.2   Polychromatic Radiation

If an optical source is polychromatic with a spectral radiant power distribution $\Phi_\lambda(\lambda)$, its luminous flux Φ (in lumens) is calculated by taking the weighted sum of $\Phi_\lambda(\lambda)$ (in watts per meter) and $V(\lambda)$ and multiplying the sum by 683. Thus,

$$\Phi = 683 \int_{\lambda_1}^{\lambda_2} \Phi_\lambda(\lambda) \cdot V(\lambda) d\lambda \qquad (4.3)$$

where λ_1 and λ_2 are the lower and upper wavelength boundaries of the optical source spectrum respectively. Equation (4.3) is essentially the same as Equation (2.12) in Chapter 2.

4.2.4   General Photometric Measurements

As seen in the preceding sections, radiometry is essential to photometry. Thus, instruments that perform photometric measurements must also perform radiometric measurements and have the ability to convert to a photometric domain appropriately under various ambient conditions. In Table 4.1, items 1 through 8 (with the exception of item 6) are photometric quantities, which can be measured using various instruments that are commercially available. The measurement aspects of these quantities along with some exemplary instruments are discussed next.

4.2.4.1   Luminous Flux, Φ

Luminous flux is the fundamental photometric quantity that corresponds to the amount of electromagnetic radiation emitted by an optical source, as seen by the average human eye. It is a scalar quantity that is always measured in units of lumen. As seen in Section 4.2.3, it is quantified by measuring the spectral radiant power, which is then weighted with the average human eye spectral response function known as the "luminosity function," $V(\lambda)$. Because the human eye is variably sensitive over the visible spectrum as seen in the $V(\lambda)$ function, monochromatic light sources of different colors yield substantially varying lumen outputs for the same amount of radiant power. For monochromatic and polychromatic electric light sources, the ratio of total output luminous flux to total input electrical power gives the luminous efficacy. Luminous efficacy is an important parameter relating to energy efficiency, particularly when white-light illumination lamps are being compared.

Total luminous flux of a light source is most straightforwardly measured using an enclosed apparatus with detector and source ports or stations known as an *integrating sphere* [70]. Such integrating spheres are designed to detect the emitter's total radiant power and all the spectral radiant power components by homogeneously distributing its optical radiation through numerous Lambertian reflections off the sphere's inner surface. Thus, they allow total flux measurements for many different types of directional light sources, including LED and fluorescent lamps. Several ideal concepts are employed for this apparatus and it is therefore important to ensure that such a constructed sphere satisfies the idealities within very good approximations. Among these include the inner surface reflections, which should be close to the Lambertian kind and reflectance nearly wavelength independent; further, the detector should be insensitive to the spatial and polarization characteristics of incident radiation. Another critical requirement is the diameter of the sphere, which must be several times larger than the overall size of the light source sample as well as the size of the entrance and exit ports.

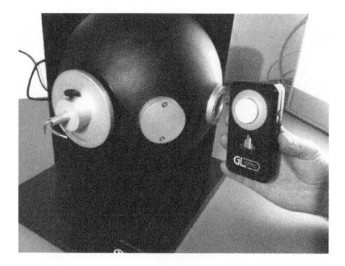

Figure 4.2. Integrating sphere GL OptiSphere 205 (left) and GL Spectis 1.0, a hand-held light meter from GL Optic (a subsidiary of Just NormLicht GmbH in Germany). GL OptiSphere 205 is a compact integrating sphere that is convenient and versatile for making numerous fast measurements of suitable light sources including various types of LED lamps. (Photo courtesy of GL Optic GmbH.)

Integrating spheres are rather sophisticated measurement instruments that may require complex calibration, operation, and maintenance. They are also capable of measuring many other photometric, radiometric, and colorimetric measurements.

A number of manufacturers, including GL Optic, Gamma Scientific, Labsphere, Instrument Systems, and Gigahertz-Optik, are currently offering integrating spheres in various sizes for luminous flux measurements of monochromatic and white light sources [71–74]. Figure 4.2 shows a compact integrating sphere by GL Optic that is able to measure total luminous flux and radiometric and colorimetric characteristics of various light sources including small LED lamps. Total flux is an important parameter for qualifying LEDs at various production line stages, including single and arrayed, unpackaged and packaged, and with secondary optics in the final assembly. The flux measurement must be practical and repeatable at all these stages, when relevant.

4.2.4.2 Luminance, L

Luminance is an inherent property for a light source that remains unchanged with respect to spatial dimensions as long as the source uniformly radiates light from its surface. It is the same as luminous intensity per unit area and therefore quantifies how much light is present in a particular direction per unit surface area enclosing a unit volume. In other words, it provides the source's inherent brightness information. It is measured in lumens per steradians per square meter ($lm/sr/m^2$) or candelas per square meter (cd/m^2)—also known as *nits*. It is an important quantity for certain lamps, such as those for automotive headlights and projection lamps, as well as electronic displays and computer monitors.

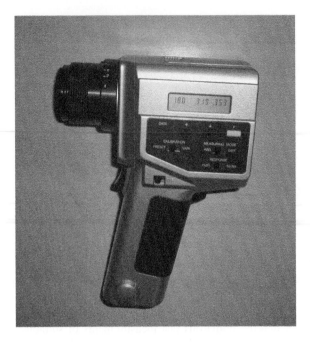

Figure 4.3. Luminance and color meter CS-100A from Konica Minolta (Japan) that measures luminance of optical spot sizes with diameters as small as 1.3 mm with close-up lenses. It also measures (x,y) color coordinates. The higher series, CS-200A, measures spot sizes as low as 0.3 mm.

Careful measurements of luminance can reveal the source's brightness uniformity and discover if there are any dark spots within a luminaire that cannot be detected easily with a naked eye. Although it is widely used in the display industry, it is not considered a very common parameter in the general lighting industry. However, LED lamps' and luminaires' users and specifiers would greatly benefit if manufacturers included this parameter as well as its spatial profile in the specification sheet for several applications. Konica Minolta has various luminance meters that can measure luminance of light sources with various spot diameters, some of which are well suited for single LED emitters [75]. Figure 4.3 shows Konica Minolta's CS-100A, which is a hand-held meter that can measure luminance of small spot diameters with close-up lenses and color properties of light sources.

4.2.4.3 Illuminance, E_v^{*}

Lighting designers and architects routinely measure illuminance of lamps and luminaires to determine the amount of luminous flux incident on a surface that is some distance away from a single or multiple light sources. They seek certain illuminance values on certain planes to ensure that proper illumination is achieved for a desired application. Because illumination planning also includes selecting color properties, integrated devices are now available to measure illuminance and

---

* The subscript "v" is used to denote "visible" radiation.

Figure 4.4. Illuminance spectrophotometer CL-500A from Konica Minolta (Japan); it measures illuminance, color parameters, and spectral properties of light sources.

color characteristics. Konica Minolta's CL-500A, shown in Figure 4.4, is such a device that measures illuminance in lux (lumens per square meter) or footcandles, and (x,y) and (u',v') CIE color coordinates, color temperature, and spectral radiant power distribution of light [76].

4.2.4.4   Luminous Exitance, M

Luminous exitance quantifies the luminous flux per area emitted or reflected *from* a surface. In the preceding section, illuminance is described as the quantity that provides how much flux is incident *on* a surface. While illuminance is predominantly used for illumination applications, luminous exitance is its analogue for display applications. As one would expect, both quantities use the same unit (i.e., lumens per square meter). However, the abbreviated term, *lux* (lx is the unit), is used *only* for illuminance.

4.2.4.5   Flux Distribution

Spatial flux distribution (SFD) is expected to become a more important lighting parameter, at least from design and measurement perspectives, as LED lamps become more common in the market. This is because, unlike traditional sources, LEDs are small, directional, and very bright as discrete emitters. Therefore, using an array of them in a luminaire without proper secondary optics typically produces substantially nonuniform luminous intensity distribution (LID) and SFD within the illuminated region of interest.

LID and SFD are analog parameters, where the former relates to the light source and is its inherent property and the latter relates to the optical power distribution the light source generates in space. SFD is essentially the same as an illumination map.

In contrast to LED sources, incandescent and fluorescent lamps by nature have larger, curvaceous, and continuous emitter surfaces that produce more uniform luminous intensity and SFD over wider spatial regions. Because SFD and LID are difficult, laborious, and expensive parameters to measure, lamp and lighting designers are usually content if they play a less critical or an optional role. In such cases, they may proceed instead with utilizing prior data of similar products that provide a good approximation.

Spatial flux distribution describes how the lumen output power is profiled over space. SFD and LID characteristics are measured using goniometric or gonio-photometric setups. The difference between a goniometer and a goniophotom-eter is, of course, that the latter is determined from the former after the $V(\lambda)$ adjustment. Luminous intensity distribution is commonly measured in phi-theta angular coordinates, which can yield the desired SFD information for a particular region of interest. Angular LID data can also be summed to calculate the total luminous flux, Φ, for homogenous and reasonably regular shaped lamps.

The LID data of light sources can be obtained from several types of gonio-photometer measurement systems. These systems are based on moving the light source, moving a mirror, or moving a detector that may be a sophisticated imag-ing camera to capture luminous flux, intensity, or luminance profile of the source. Figure 4.5 shows a goniometric setup schematic where the apparatus's detector is moved about the sample that remains fixed. In a similar type of configuration, the angular LID data for lamps can be obtained from a setup that uses two rotational movements of the detector, each supporting the *phi* or *theta* angular orientation of the test lamp to the detector. The detector diameter and the distance between the

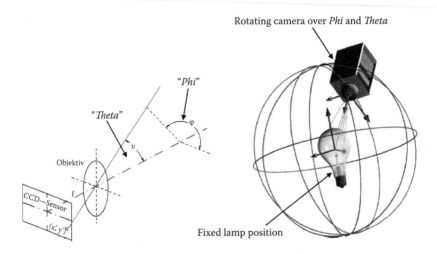

Figure 4.5. Schematic diagram showing the measuring principle of a near-field goniophotometer from Techno Team Bildverarbeitung GmbH (Germany). The detector is a high-resolution imaging camera that is continuously moved over the shown spherical surfaces that surround the fixed lamp at the center. (Figure courtesy of Techno Team Bildverarbeitung GmbH.)

lamp sample and the detector form the solid angle for which the angular LID is measured and repeated over various *phi* and *theta* domains.

Companies such as Techno Team Bildverarbeitung GmbH and Instrument Systems offer goniophotometers that are suitable for various light sources, including LEDs, in the market [77,78]. The RiGO-801 goniophotometer system by Techno Team enables near-field SFD measurements for lamps and luminaires of various sizes (including large commercial fluorescent lamps) with high precision by utilizing a luminance measuring camera system on a positioning stage [77]. This is shown in Figure 4.6. TechnoTeam also has smaller goniometers for measuring small LED lamps and modules.

The commercially available goniophotometers generate LID data that can be converted to various standard file formats for the lighting industry; some can also be converted to ray files compatible with popular ray-optic design tools. For instance, the RiGO-801 system generates LID data that can be converted to such data files as EULUM-DAT, TM14, IES, and Calculux using the photometric database LUMCat. Using Techno Team's free conversion software, their LID data can also be converted to ray files compatible with ASAP, SPEOS, Lighttools, and Zemax, which are well-known design tools in the lighting industry [79].

Photograph courtesy of Techno Team Bildverarbeitung
GmbH (Germany)

Figure 4.6. The RiGO-801 near-field goniophotometer for measuring LID of lamps and luminaires of various sizes. The lamp or luminaire remains fixed while the swivel-mounted goniometer arms holding the detectors sweep along the horizontal and vertical directions about the light source sample.

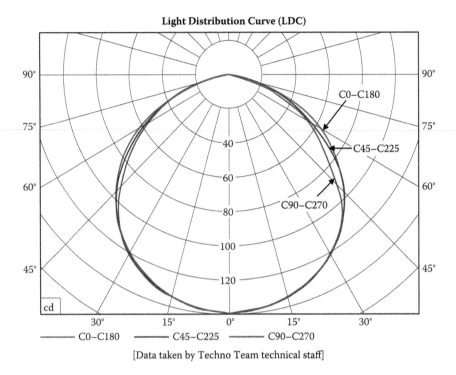

Light Distribution Curve (LDC)

[Data taken by Techno Team technical staff]

Figure 4.7. The light distribution curve (LDC) of LED-S4 sample measured using a RiGO-801 goniometer. This LDC plot is equivalent to the LUMCat data presentation of LID in polar coordinates commonly used in the lighting industry.

Figure 4.7 shows the measured light distribution curve (LDC) of an LED sample, LED-S4, which will be described later in the chapter. The measurement was performed using a RiGO-801 system. This plot is equivalent to LUMCat-generated LDC plots of .IES files [80].

What makes goniophotometric measurements laborious and expensive is that the system requires mechanical drives to move and pause at many locations around the detector or light source in order to take measurements at small angular steps. Therefore, as mentioned before, they are currently only used by a few groups in the industry, while others may obtain such measurements occasionally. Goniophotometers are primarily used when SFD and LID data become essential for acute qualification of a luminaire light ray profile—for example, in street and automotive lighting where the light must be directed over a certain region for optimal illumination.

4.2.4.6 Luminous Efficacy and Luminous Efficiency

Although luminous efficacy and luminaire efficiency are not directly measured with an apparatus, they can be easily calculated from certain measured quantities. As discussed previously, using a suitable integrating sphere, a lamp's total output flux, Φ_{LAMP}, may be measured in lumens for a certain time-averaged input electrical power, W_{av}, in watts. Efficacy is the quotient, Φ_{LAMP}/W_{av}, in units of lumens per watt (lm/W).

If a luminaire is constructed by inserting a lamp in some fixture, the total luminaire output flux, $\Phi_{LUMINAIRE}$, as well as the lamp's flux, Φ_{LAMP}, can be measured using a suitable integrating sphere. Luminous efficiency is the quotient, $\Phi_{LUMINAIRE}/\Phi_{LAMP}$, multiplied by 100.

4.2.5   General Colorimetric Measurements

Items 8 through 10 in Table 4.1 are colorimetric quantities of light sources that can be measured using a number of commercially available instruments. As does photometry, colorimetry defines quantities perceived by the human eye. The foundation of colorimetry is formed by establishing methods of quantifying physiological color perception caused by the spectral color stimulus properties of the average human eye. When the human eye perceives the environment and its objects illuminated in light, the chromatic or achromatic constituents of reflected light rays reaching the eye appear as "color" through the eye's physiological sensation. This manifestation is described by such chromatic specifications as purple, blue, green, magenta, red, brown, orange, yellow, and others; by achromatic descriptions such as white, gray, black, etc.; or by some combination thereof. They are further described by level of brightness and vibrancy. The color being perceived depends on the spectral distribution of the eye's color stimulus as well as the physical nature of the stimulus and its surroundings, such as its relative position, shape, and size. In addition, the observer's experience and visual system's adaptation play a role in determining what color is perceived.

The human eye perceives monochromatic light of different wavelengths as different colors within the visible spectrum. The perceived color is quantified by calculating the internal eye response to the external spectral color stimulus function, which is the light source's spectral radiant power distribution function, $\Phi_\lambda(\lambda)$, as explained in Equation 4.3.

The human eye also distinguishes object illumination from white light composed of a particular set of wavelengths as being different from that generated by white light composed of another set of wavelengths. Thus, the same object appears to have different color shades when it is illuminated with light sources of varying spectral color properties. The object's color is quantified by calculating the internal eye response to the external spectral color stimulus function, which is now the product of $\Phi_\lambda(\lambda)$ (the light source's spectral radiant power distribution) and the object's spectral reflectance distribution, $R_\lambda(\lambda)$, or its spectral transmittance distribution, $T_\lambda(\lambda)$.

4.2.5.1   The CIE Standard Illuminants

Since an object's color depends on how it is illuminated using a light source, the classification of the object's color requires characterizing that light source based on a certain reference light source. The CIE therefore defined the colorimetric standards for certain reference light sources (i.e., illuminants). There are primarily two such illuminants: (1) the CIE *Standard Illuminant A,* which is defined by a Planckian blackbody radiator at a correlated color temperature (CCT) of 2856 K, and (2) the CIE *Standard Illuminant D65,* which represents average daylight with a CCT of 6500 K [81].

4.2.5.2   The CIE Standard Color Spaces

The CIE also created certain mathematically defined color spaces to quantify color perception. These resulted from the experiments by W. David Wright and John Guild in the 1920s and 1930s, which combined red, green, and blue (RGB) lights to produce single colors in the visible spectrum [82,83]. Their data led to the generation of the standardized RGB color-matching functions, which were then converted into the CIE 1931 *XYZ* color-matching functions that subsequently formed the corresponding color space.

The human eye photoreceptors, or cone cells, have been shown to have sensitivity peaks in red, green, and blue wavelength regions in the presence of medium- and high-brightness ambient light (i.e., there are three primary color stimuli for photopic vision). Therefore, in principle, all colors may be expressed by means of some appropriate *tristimulus* value representation. Color spaces, including the CIE 1931 *XYZ* and CIE 1964, associate tristimulus values with colors and thereby provide the means to quantify color properties of objects and light sources. The interested reader is encouraged to learn more about color spaces and matching functions in relation to describing and quantifying colors from numerous available publications including those from CIE [84,85].

4.2.5.3   The CIE Standard Chromaticity Diagrams

The CIE 1931 *XYZ* color space in three dimensions (3-D) provides *XYZ* tristimulus values that represent all possible color perceptions. In this space, *Y* provides the luminance value, and *X* and *Z* are some appropriate derivative parameters of the tristimulus colors. A two-dimensional (2-D) representation of this, known as the CIE 1931 (*x,y*) chromaticity diagram on a plane (see Figure 1.2 in Chapter 1), is sufficient for most applications. The *x* and *y* coordinates of this diagram are calculated from a projection of the *X, Y, Z* values as the following [85]:

$$x = \frac{X}{X+Y+Z} \text{ and } y = \frac{Y}{X+Y+Z}. \tag{4.4}$$

The CIE 1931 (*x,y*) is widely used around the world. However, it has a significant disadvantage because the geometric distance between two coordinates within the plane does not correspond very well to the perceived color difference between the two points due to the nonlinear color characteristics present in the (*x,y*) representation. As a result, in 1976, the CIE introduced the uniform (*u′, v′*) chromaticity scale (UCS) diagram, where the coordinates are defined by

$$u' = \frac{4X}{X+15Y+3Z} \text{ and } v' = \frac{9Y}{X+15Y+3Z}. \tag{4.5}$$

The weighted transformation in Equation (4.5) is employed to counteract some of the nonlinear characteristics that would otherwise be present in the projected planar space representation of the *XYZ* space. While this (u', v') scale still fails to provide a strict linear correspondence between geometric distances within the plane, the discrepancies are much less than those from the CIE 1931 (x,y) diagram.

As mentioned in Section 4.2.4.3, several illuminance meters have the capability to measure (x,y) and (u', v') color coordinates of light sources. Color meters that span the wide visible spectrum are calibrated based on a CIE tristimulus reference standard that uses a source of known CCT. Calibrations of CCT, illuminance, or luminance may all be utilized depending on the color meter's capability. Figures 4.8(a) and (b) respectively show the "*xy*" and "*uv*" data taken for a 40 W incandescent lamp using a Konica Minolta CL-500A. Note that although the (x,y) and (u', v') color coordinates have different values in their corresponding color space as expected from Equations (4.4) and (4.5), both coordinates fall on the Planckian blackbody locus line—as they must because the sample is an incandescent lamp.

The additional color properties of a light source are the color rendering index (CRI) and CCT (items 10 and 11 in Table 4.1), which can be measured also using a Konica Minolta CL-500A. These quantities were described in Chapter 1. Examples of their measurements will be presented in the next section, where the color properties of LED lamps will be compared with those from incandescent and compact fluorescent lamps (CFLs). Sections 4.2.4 and 4.2.5 complete the

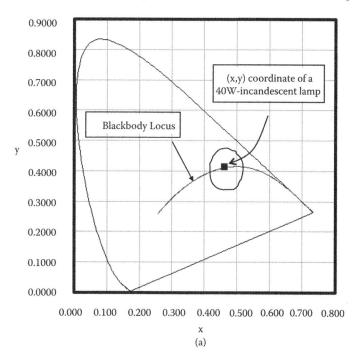

Figure 4.8. Color coordinates measured for a 40 W incandescent lamp (sample INC-S1) using a Konica Minolta CL-500A meter: (a) (x,y) coordinates of the CIE 1931 diagram; (b) (u',v') coordinates of the CIE 1976 UCS diagram.

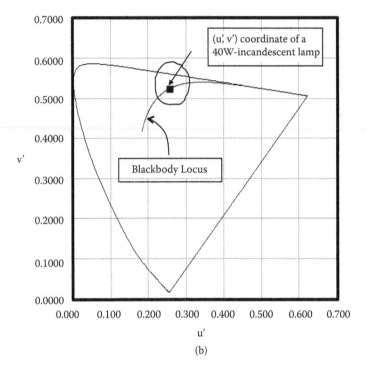

(b)

Figure 4.8. (continued) Color coordinates measured for a 40 W incandescent lamp (sample INC-S1) using a Konica Minolta CL-500A meter: (a) (*x,y*) coordinates of the CIE 1931 diagram; (b) (*u′,v′*) coordinates of the CIE 1976 UCS diagram.

basic discussions on primary photometric and colorimetric parameter characterization of light sources used in general and related lighting applications.

4.3   Application of Standard Photometry and Colorimetry to LED Lamps

Most lamps and luminaires are traditionally evaluated against the basic lighting metrics provided in Table 4.1. LED lamps should be no exception. It is important to note that *all* of these metrics should be measured *for all ranges of interests* when characterizing a lamp for its own merit or comparing it against its counterparts. LED engineers and developers must be careful to not always compare parameters characterized by single values for an absolute determination of *superior versus inferior* lamps; rather, when appropriate, they should use meaningful parameter ranges to describe and compare certain performance specific to applications. For example, although illuminance and luminous efficiency are both characterized by single values, they may be location dependent if lamp sources of different shapes and positions are used for comparison. Similarly, total luminous flux and efficacy are inadequate single values if used without other parameters such as flux distribution. Single-value comparisons do not particularly suffice for ambient or broad-space lighting applications. Furthermore, comparing color rendering capabilities of two light sources with a traditional single-valued CRI is often inadequate.

These challenges are especially pronounced when LED lamps are compared to incandescent and fluorescent lamps, but are less crucial when incandescent lamps are compared to compact fluorescent lamps. This is true because typically, emitter shapes and spectral content of LEDs are very different compared to those from their incumbent counterparts. The different spectral content in different types of lamp sources, such as in LED and incandescent lamps, affects color rendering properties in ways that are not properly translated by the single CRI value defined in a traditional manner. Therefore, for the current general CRI definition to remain as an adequate comparison index is being challenged, and the topic of redefining CRI or broadening the description of comparative color quality is now of interest to some in the industry. At present, a debate is taking place to use a different metric, known as color quality scale (CQS), for describing color rendering properties of LED lamps [86].

4.3.1 Lamp Measurements and Comparisons

Let us now evaluate several LED, incandescent, and compact fluorescent lamp samples. These lamps are all commercially available; most are available in retail stores and the others can be ordered through special vendors. As mentioned earlier, it is important to distinguish applications when evaluating or comparing lamps. Therefore, here we evaluate three categories: (1) ambient lamp, (2) task lamp, and (3) decoration lamp.

4.3.1.1 Ambient Lamp Measurements

Ambient lamps are generally used to suit certain environments such as dining, ballroom gatherings, or some sort of exhibition. For these indoor events, which are usually evening occasions, the illumination must be fairly omnidirectional and provide high CRI, warm CCT, and certain minimum illuminance levels in desired areas from individual lamps. Depending on the lamp and luminaire configurations, the number of individual lamps should be optimum for a given space; that is, utilization of too many or too few lamps or luminaires should be avoided to achieve some optimum illuminance that may be required to recognize and appreciate objects of interest in a desired manner. Generally, the optimum illuminance requirements are needed near the middle of the room or at other gathering locations, sometimes over an appreciable range of vertical dimension rather than on one incident plane at some height. Corners typically need not have high illuminance, but must have some contrasted visibility.

Traditional incandescent or improved CFL replacement lamps are suited for these types of ambient illumination applications. Some LED lamp replacements intended to be used in similar environments are now being offered in retail stores. We now compare the photometric and colorimetric properties of three such LED lamps against three incandescent and three CFL bulbs. The LED bulb samples are referred to as LED-S1, LED-S2, and LED-S3; the incandescent lamp samples are denoted as INC-S1, INC-S2, INC-S3; and the CFL samples are referred to as CFL-S1, CFL-S2, and CFL-S3. In Figure 4.9, the measured CRIs for the three samples of each lamp type are shown.

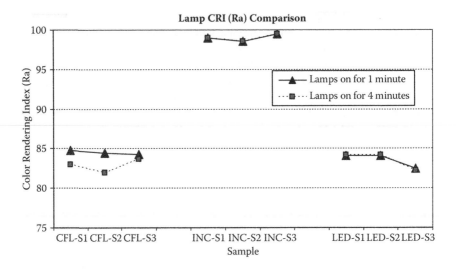

Figure 4.9. The measured CRI (Ra) using Konica Minolta CL-500A for three samples of CFL, incandescent, and LED lamps are shown for two test conditions: (1) data taken after lamps stayed on for 1 minute, and (2) data taken after lamps stayed on for 4 minutes.

It is important to note that the CRI degraded for most CFLs as the lamps remained on for a longer period of time, while the CRI remained the same for the LED and incandescent lamps. Both CFL and LED lamps have lower CRI values than what would be desired for the evening events mentioned before.

However, as discussed earlier, a single CRI value is often insufficient to make a color rendering comparison over a broad spectrum. The current general standard for CRI is **Ra**, which defined as the average of eight sample color ratios referenced to an ideal illuminant, the choice of which depends on whether the sample light source's CCT is above or below 5000 K [87,88]. Another seven samples provide supplementary information on color rendering properties of the light source such as its saturation levels and color comparisons against well-known objects [89]. Although more elaborate and complete CRI definitions, such as the **R96a** [90], are available, **Ra** is still widely used. Despite the fact that **Ra** is just one number, it may be more useful to use a graphical representation of **Ra** where all 15 color samples are simultaneously shown in a polar plot, as depicted in Figure 4.10.

When comparing CRIs of two light sources, it is helpful to observe two polar plots side by side to see which one better fills the circle. For example, in Figure 4.11, the polar plots of CRIs from CFL-S3 and LED-S3 are shown, which provides a more effective method for judging how well various common colors will be rendered by each source. The color bar plots are less crucial, particularly for experienced workers who have become familiar with the sample colors. When the number of samples to be compared is small, polar plots as in Figure 4.11 are perhaps more useful for making CRI comparisons than the presentation shown in Figure 4.9.

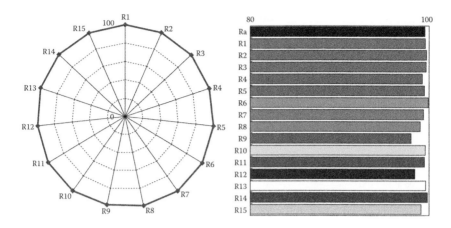

Figure 4.10. (**See color insert.**) The polar and bar plots of the measured CRI (Ra) data of INC-S3. The color bar plot shows the color and strength of each R_i sample, while the polar plot illustrates that a high CRI would amount to filling the approximate polar circle as closely as possible. The data were taken with Konica Minolta CL-500A.

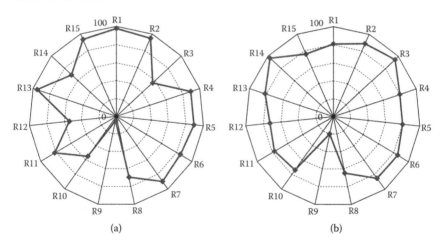

Figure 4.11. The measured CRI (Ra) plotted for two light sources: (a) CFL-S3, and (b) LED-S3, after the lamps were on for 4 minutes. Observing polar plots of two light sources side by side should give a more comprehensive weighing ability when comparing their CRIs. The data were taken with a Konica Minolta CL-500A meter.

While the CRI of a light source indicates how well the viewers will see colors, it does not indicate the apparent color of the light source. The CCT of the light source provides that information. For ambient lighting desired for dining and exhibiting, CCTs must be in the range between 2700 and 3000 K. We now look at the CCT performance of the samples under discussion.

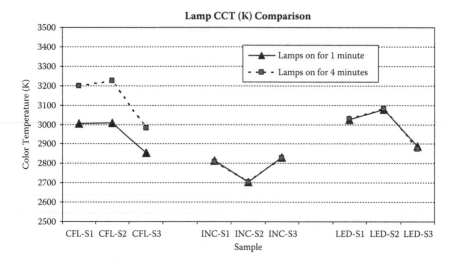

Figure 4.12. The measured CCTs using Konica Minolta CL-500A for three samples of CFL, incandescent, and LED lamps are shown for two test conditions: (1) data taken after lamps stayed on for 1 minute, and (2) data taken after lamps stayed on for 4 minutes.

Figure 4.12 shows the measured CCT for the three samples of each lamp type. It is interesting to note that the CCT increased for all the CFL samples as the lamps remained on for a longer period of time. In contrast, the CCT remained nearly the same for LED and incandescent lamps. As expected, the incandescent samples have the warmest CCT; however, the CFL and LED lamp samples here show reasonably low CCT values and they would be considered acceptable for many types of evening events. If the CFL samples turned significantly cooler in color temperature after being on for some time, then they would likely not be acceptable for such uses.

Ambient lighting designers will require certain lumen output over some specified regions in a room. As discussed in the previous section, this is where quantifying SFD would be important; this requires goniophotometric measurements. However, at a minimum, lighting designers would like to ensure that certain illuminance levels are met. If a lamp's LID properties are fairly well known, measuring illuminance at certain planes provides reasonably good information on how certain other places of interests would be illuminated because SFD can be extrapolated from the LID data. For a relative comparison, illuminance data were taken for all nine samples at 17 in. below and 5 in. across from the lamp's center. These data along with the CRI and CCT data presented in Figures 4.9 and 4.12 are provided in Table 4.3.

The photographs of the CFL, incandescent, and LED samples of Table 4.3 are shown in Figures 4.13, 4.14, and 4.15 respectively. Before wrapping up the discussion on ambient lighting sample measurements, it is important to elaborate the measurement conditions and sample descriptions further. Photometric and

Table 4.3. CRI, CCT, and Illuminance Data for CFL, Incandescent, and LED Samples Using Konica Minolta CL-500A

Data Taken July 12, 2012		CRI Data (Ra)		CCT Data (K)		Illuminance (lux)	
Rated Sample Wattage (W)	Sample	1 min	4 min	1 min	4 min	Incident Plane from Bulb: Down: 17 in.; Over: 5 in.	
						1 min	4 min
9	CFL-S1	85	83	3005	3200	443.00	413.20
9	CFL-S2	84	82	3009	3225	487.07	438.75
11	CFL-S3	84	84	2856	2982	854.13	739.19
60	INC-S1	99	99	2814	2806	1111.05	1115.90
60	INC-S2	99	99	2704	2702	910.58	910.60
75	INC-S3	100	100	2831	2828	1638.57	1636.43
13.5	LED-S1	84	84	3026	3029	1215.46	1208.89
7.5	LED-S2	84	84	3079	3083	902.49	900.92
8	LED-S3	83	82	2887	2873	692.31	679.34

Figure 4.13. A photograph of the three CFL samples used in the lamp comparison study in this section. These lamps are currently available in major retail stores in the United States.

colorimetric evaluations of most light sources including LEDs are affected by operating temperature. All measurements presented here were performed at a steady room temperature near 22°C. Measurements were performed only after stable readings for all parameters were achieved. Each data value is an average of 10 points taken at 1-second intervals.

As seen in Figure 4.13, the first two CFL samples presented here are not the conventional ones that have twisted shapes and are usually somewhat bigger in size. Smaller sizes and shapes that lack twists affect the overall tube length, which negatively impacts the performance of CFLs. These may be part of the reasons for the sub-par performance observed in the CFL data presented here. The bigger and more twisted shapes provide longer interaction length for generating

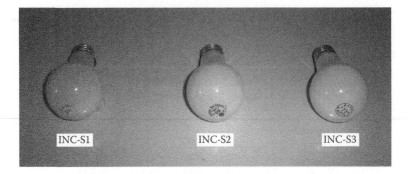

Figure 4.14. A photograph of the three incandescent lamp samples used in the lamp comparison study in this section. These lamps are presently available in retail stores in the United States.

Figure 4.15. A photograph of the three LED lamp samples used in the lamp comparison study in this section. These lamps are currently available in several major retail stores in the United States as of this writing.

fluorescence light output more effectively and therefore better performance is expected. Although CFL-S3 has a twisted lamp inside the outer glass enclosure, its ability to dim may have compromised its lighting performances presented here. While more tests need to be done to confirm these and additional conclusions, lighting designers and end users would be wise not to treat all CFLs in the market equally with regard to their performances.

4.3.1.2　Task Lamp Measurements

Task lamps are becoming increasingly popular as more people become employed throughout the world and as people utilize more personal computers of various kinds to perform everyday duties in their lives. Task lamps allow for brighter illumination over some desired space needed to perform a task where a certain amount of visibility is required, while keeping the surrounding ambient illumination at a dimmer level.

Traditional incandescent or CFL replacement lamps are usually not suited for illuminating surfaces and smaller regions to perform typical tasks because these light sources emit light nearly omnidirectionally. Hence, lamp shades and particular types of luminaires are designed to reflect light from certain directions to provide illumination downward over a confined region. Since basic LED lamps emit light more directionally over a limited region, they are naturally better suited for many task illumination applications. Currently, LED task lamps can be ordered through a number of vendors. We now compare the photometric and colorimetric properties of an LED task lamp against a CFL lamp that has a similar electrical input wattage rating. The LED task sample is referred to as "LED-S4" and the CFL sample is referred to as "CFL-S4." The rated wattage for the LED and CFL samples were 5 and 7 W respectively. Figures 4.16 and 4.17, respectively, show the photographs of these LED and CFL lamp samples.

Figure 4.16. A photograph of the task light sample, LED-S4. The size is referenced with a US 10-cent coin (dime).

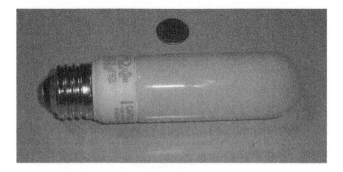

Figure 4.17. A photograph of the task light sample, CFL-S4. The size is referenced with a US 10-cent coin (dime).

Table 4.4. Measured CRI and CCT Data for CFL-S4 and LED-S4 Task Lamps

| Data Taken July 13, 2012 | | | |
Rated Sample Wattage (W)	Sample	CRI Data (Ra) Lamp on > 30 min	CCT Data (K) Lamp on > 30 min
7	CFL-S4	82	6402
5	LED-S4	89	3035

The desired colorimetric properties for task lamps usually are that they have CRI over 85 and CCT between 2,700 and 5,000, depending on whether the task is being performed during the day or at night and depending on the user's preference. The CRI and CCT of the two task lamps are shown in Table 4.4.

The CRI and CCT values for the LED task lamp sample are well within the acceptable ranges. However, they are undesirable for the CFL task lamp sample. To investigate the color quality of these two samples further, their spectral radiometric power distributions were measured. These are shown for LED-S4 and CFL-S4 in Figure 4.18 and Figure 4.19 respectively. These spectral power distributions explain the desirable and undesirable color properties of the LED and CFL samples respectively.

The photometric requirements of task lamps are that they provide a certain level of illuminance and that a specified range of illuminance levels be maintained uniformly or nearly uniformly over the spatial region of interest so that proper task visibility can be achieved. In order to quantify such properties, an illuminance mapping experiment was performed to measure the illuminance from the two lamp samples on a desk top. A lamp holder was placed on the desk top in such a manner that its base center served as the reference in the *X-Y* plane,

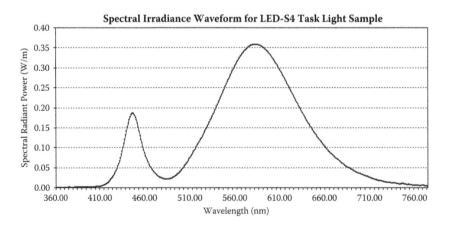

Figure 4.18. The measured spectral radiant power distribution for sample LED-S4. Fairly broad spectral distribution about the peak wavelength of 586 nm leads to a high CRI of 89 and warm color temperature of 3035 K.

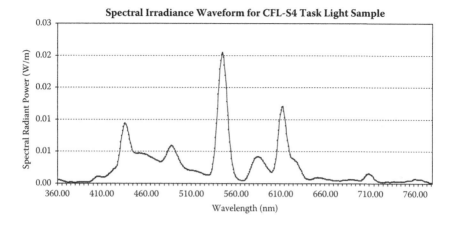

Figure 4.19. The measured spectral radiant power distribution for sample CFL-S4. Several narrow spectral distributions about red, green, and blue wavelengths lead to a poor CRI. The cool color temperature of 6402 K results from the lack of spectral power in the warm wavelength regions.

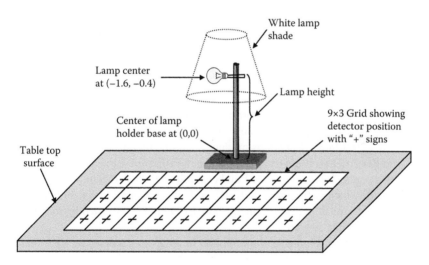

Figure 4.20. The schematic drawing of the illuminance mapping experimental setup to compare LED and CFL task lamps.

which is desk top surface. This reference is the origin (0.0, 0.0) of the X-Y coordinate system describing the desk top as shown in Figure 4.20. The X-Y coordinate system is a (9 × 3) rectangular grid where the X- and Y-grid units measure 2.5 and 6 in. respectively.

The longitudinal Z-direction signifies the height of the lamp; two sets of illuminance data were taken at two different heights for both lamp samples. The lamp holder used in the experiment was part of a desk lamp type of luminaire that included a white shade to direct light downward from omnidirectional sources. The (x,y) position where the lamp samples are attached to the luminaire is (–1.6, –0.4); that is, they were 4 in. in the negative X-direction and 2.4 in. in the negative Y-direction from the origin, as shown in Figure 4.20. The two different CFL lamp heights used in the experiment were 17.0 and 20.5 in.; the two corresponding LED lamp heights were 0.5 in. higher (i.e., 17.5 and 21 in.), due to the special mounting scheme needed for the non-Edison type connector base.

The illuminance data for CFL-S4 and LED-S4 were taken by positioning the Konica Minolta CL-500A at the 27 grid nodes shown in Figure 4.20 as "+" signs. The data for the two lamps at the lower heights are plotted in Figure 4.21 and the data for the higher heights are plotted in Figure 4.22.

The illuminance datasets show that the LED sample is significantly better in terms of providing higher luminance over a broader area on the desk-top surface. In fact, the LED sample primarily meets the minimum and recommended illuminance requirements of 50 foot-candles for task visibility set forth by GSA and underwritten by the US Department of Energy (DOE) [91]. However, this requirement is not met by the CFL sample used in this experiment, which had an even higher input wattage rating. Since the lamp heights and illuminated area used in the experiment are fairly typical for many practical tasks, one can conclude that LED lights such as LED-S4 are quite suitable for task illumination applications. The results in this experiment have quantitatively shown that LED task lamps offer better performance than their fluorescent counterparts, which of course are superior to their incandescent counterparts in terms of energy efficiency. It is therefore not surprising that, in recent years, most lighting designers and end users have expressed preference toward LED task lighting.

Utilization of LED light sources as task lamps is also embraced by many despite their undesirable glare properties because users need not see them directly since the lamps would most often be placed within a shade as shown in Figure 4.20. Many LED lamps produce a significant amount of glare because they use flat LED emitters on a plane as seen in Figure 4.16. Such an arrangement of discrete LEDs produces Lambertian-like light distribution, as seen in Figure 4.23, which is the measured LID performance of LED-S4. In Chapter 6 and Chapter 7, we shall discuss further why LED lamps with such light distribution are susceptible to glare.

The LID data in Figure 4.23 are plotted in terms of candelas in XYZ spatial coordinates. The sample was oriented at the center of the XYZ coordinate system, facing outward in the positive Y-direction with its LED emitter plane parallel to the XZ plane. The figure shows that all the light emitted from the LED lamp concentrates in front of the lamp, resembling a slightly modified 3-D Lambertian.

4.3.1.2.1 Decoration Lamp Measurements   LED lamps of various sizes and colors are widely employed for decorative applications. These fall into the viewing category, rather than the illumination category. Therefore, the lamp requirements usually are not very stringent with respect to CRI, CCT, and lumen output requirements. Because discrete LED lamps can be packaged into arrays in plastic casings and can be electronically controlled, they are well suited for decorative lighting applications.

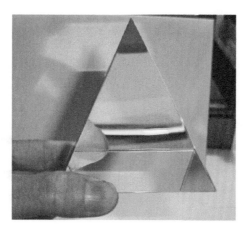

Color Figure 1.1. White light from a compact fluorescent lamp is reflected from a shiny object and the light rays are captured by a camera after passing through a prism. The rays passing through the prism in this picture show the range of individual colors within the broad white spectrum of the fluorescent lamp.

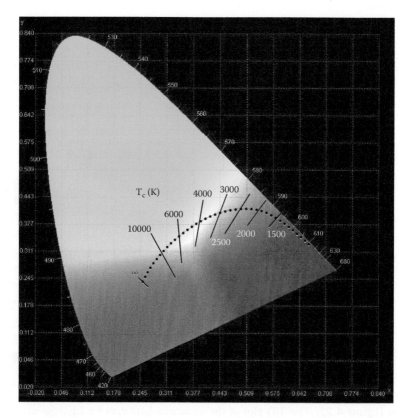

Color Figure 1.2. An indicative diagram of the CIE 1931 (*x,y*) chromaticity space generated from a spectrometer by GL Optic GmbH using their GL SpectroSoft software. The dotted line here illustrates the black-body locus, which shows various points of constant CCT represented by the intersecting lines through it.

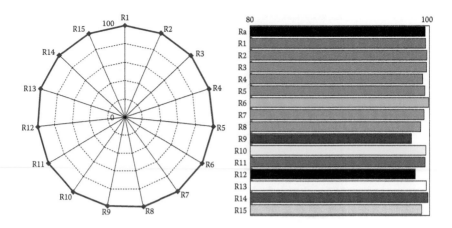

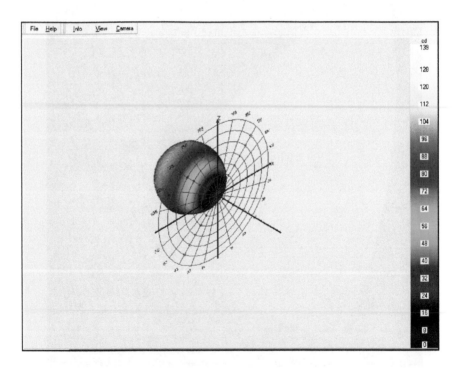

Color Figure 4.10. The polar and bar plots of the measured CRI (Ra) data of INC-S3. The color bar plot shows the color and strength of each R_i sample, while the polar plot illustrates that a high CRI would amount to filling the approximate polar circle as closely as possible. The data were taken with Konica Minolta CL-500A.

Color Figure 4.23. The 3-D plot of the measured LID data for LED-S4 using a RiGO-801 system. The graph shows the lamp's luminous intensity distribution in **XYZ** spatial coordinates, as well as the angular-space grid in the **XZ** plane. The data from the LED lamp show a near Lambertian profile with all of the light distribution in the hemisphere in front of the lamp. (Data taken by Techno Team staff.)

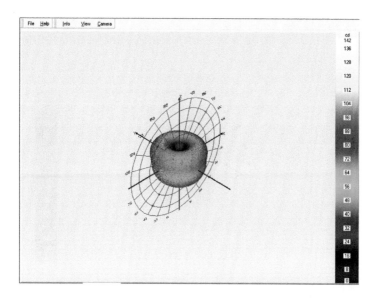

Color Figure 6.34. The 3-D plot of the measured LID data for LED-S1 (12 W) using a RiGO-801 system. The graph shows the lamp's luminous intensity distribution in **XYZ** coordinates (in position space), which also shows the angular space grid in the **XZ** plane. The LID data of the LED lamp is largely asymmetric with much of its light distribution remaining in the lower hemisphere where the z-values are all negative. (Data taken by Techno Team staff.)

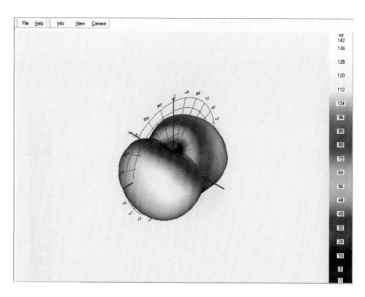

Color Figure 6.35. The 3-D plot of the measured LID data for a standard 90 W incandescent lamp using a RiGO-801 system. The graph shows the lamp's position space LID data in **XYZ** coordinates, which also shows the angular space grid in the **XZ** plane. The lamp's LID is spread quite symmetrically over 4π(sr) in both upper and lower hemispheres, showing gradual LID variation over a broad range of values. (Data taken by Techno Team staff.)

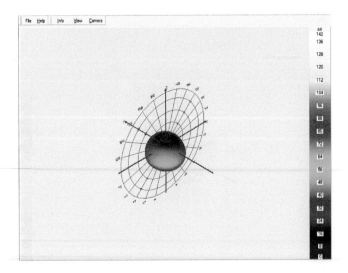

Color Figure 6.36. The 3-D plot of the measured LID data for an 11 W compact fluorescent lamp using a RiGO-801 system. The graph shows the lamp's position space LID in **XYZ** coordinates, which also shows the angular space grid in the **XZ** plane. This lamp's LID values are spread somewhat more uniformly over $4\pi(\text{sr})$ compared to that of LED-S1 in Figure 6.34. However, this CFL lamp shows higher LID strengths in the lower hemisphere. Its LID variation covers a more limited range compared to that of the incandescent lamp as observed in Figure 6.35 because the incandescent lamp has a much higher equivalent wattage. (Data taken by Techno Team staff.)

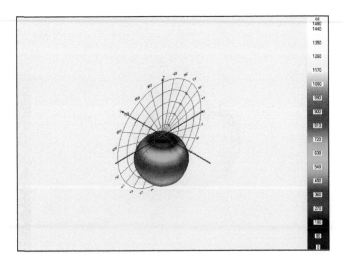

Color Figure 7.16. The measured LID data for LEDGREEN-T8-S1/S2 pair using a RiGO-801 system. The graph shows the lamp's luminous intensity distribution in **XYZ** spatial coordinates while showing the angular-space grid only in the **XZ** plane. This LID profile also resembles a near Lambertian such as that in Figure 7.15, but has much greater LID strength and distribution spread. (Data taken by Techno Team staff.)

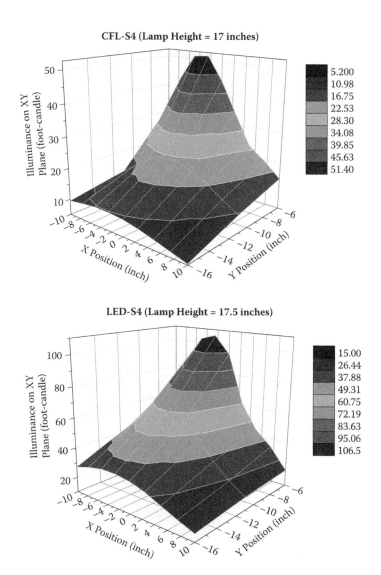

Figure 4.21. The illuminance measurement data taken on a desk-top surface for task lamps CFL-S4 and LED-S4 at lamp heights of 17 and 17.5 in. respectively. CFL-S4 provides less than half the foot-candles at the same locations on the desk surface compared to what LED-S4 provides. CFL-S4 illumination also covers a smaller region on the desk surface.

CFL-S4 (Lamp Height = 20.5 inches)

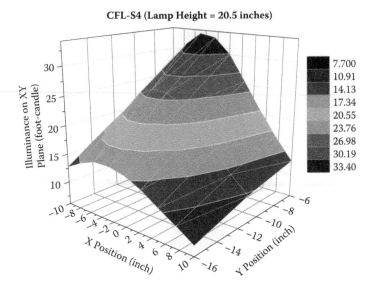

LED-S4 (Lamp Height = 21 inches)

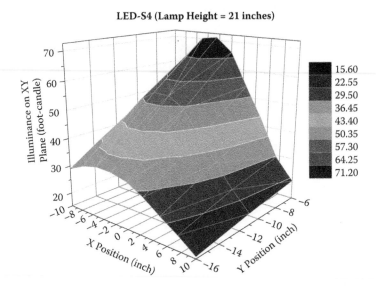

Figure 4.22. The illuminance measurement data taken on a desk-top surface for task lamps CFL-S4 and LED-S4 at lamp heights of 20.5 and 21 in. respectively. CFL-S4 provides less than half the foot-candles at the same locations on the desk surface compared to what LED-S4 provides. CFL-S4 illumination covers a similar region on the desk surface at this height.

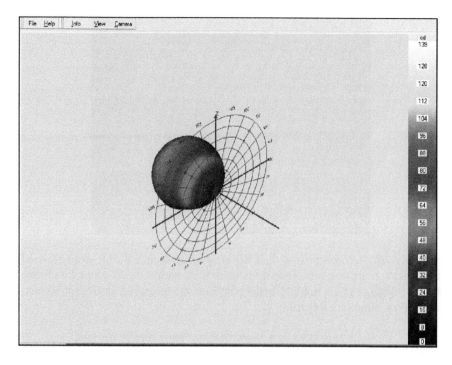

Figure 4.23. **(See color insert.)** The 3-D plot of the measured LID data for LED-S4 using a RiGO-801 system. The graph shows the lamp's luminous intensity distribution in *XYZ* spatial coordinates, as well as the angular-space grid in the *XZ* plane. The data from the LED lamp show a near Lambertian profile with all of the light distribution in the hemisphere in front of the lamp. (Data taken by Techno Team staff.)

Let us now investigate the photometric and colorimetric properties of an LED array from a commercial LED string light unit used for decorative applications. This is referred to as the "LED-Array-S1 sample." A photograph of this sample is shown in Figure 4.24.

As seen in Figure 4.24, the LED string lights are white. Although single-color LEDs are popular in many decorative applications, white LEDs are also often highly desired. Because decorative lights are directly viewed, it is recommended that the photometric parameter, luminance, be specified rather than illuminance. This is to ensure that the lamp's maximum luminance value does not exceed the comfort range of the human eye, while the minimum value remains adequate for light visibility. To quantify these parameters, 15 LEDs from the LED-Array-S1 sample (marked in Figure 4.24) were measured for luminance and illuminance. These are shown in Figures 4.25 and 4.26 respectively.

The luminance data were taken using a Konica Minolta CS-100A, with the aid of a close-up lens 110. The add-on lens allows the measurement of spot sizes as low as 1.3 mm. But because the CS-100A is a hand-held meter and the LED emitter spot sizes were nonuniform, some difficulties were encountered in accurately measuring the luminance of several LEDs in the LED-Array-S1 sample.

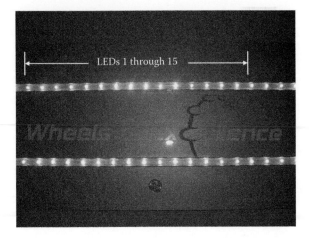

Figure 4.24. A photograph of the LED lamp array sample, LED-Array-S1, used in decoration applications. The size is referenced with a US 10-cent coin (dime). LEDs 1 through 15 are marked on the photo and are measured for their photometric and colorimetric properties.

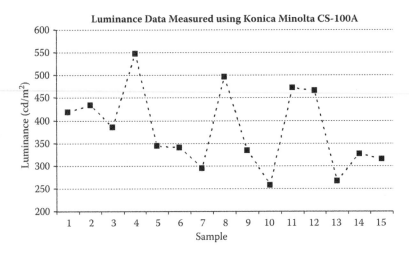

Figure 4.25. Measured luminance data for LEDs 1 through 15 from the LED-Array-S1 sample. The variation is significant, but most LEDs remain within the acceptable luminance range for viewing.

Nevertheless, clear near-field image patterns of the LEDs were seen using the CS-100A, which confirmed that the spot sizes in the 15 LEDs not only lacked uniformity, but many were also discontinuous and consisted of multiple smaller spot sizes. For such cases, luminance measurement via an apparatus such as the CS-100A would be somewhat inaccurate. A near-field, high-resolution imaging camera would be more appropriate in measuring luminance of such irregular LED emitters.

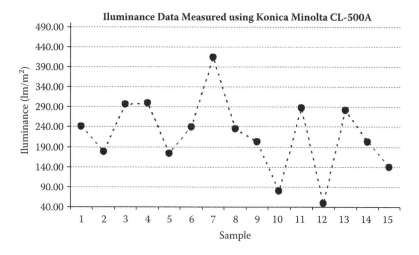

Figure 4.26. Measured illuminance data for LEDs 1 through 15 from the LED-Array-S1 sample. Although the variation is significant, it is acceptable for most viewing applications. The measurements were taken at an approximate detector height of 2 in. from the LEDs. Each LED was measured individually, covering all other LEDs.

Since it was recognized that there were some uncertainties in the luminance data shown in Figure 4.25, the illuminance data were taken for the same LEDs using a Konica Minolta CL-500A. The advantage of this instrument was that it could be securely mounted above the LED emitters and stable readings could be ensured. The significant variation in the illuminance data seen in Figure 4.26 confirmed that the LEDs in the array were in fact inherently nonuniform with respect to their size and emission characteristics. The luminance and illuminance parameters are related via Equation (1.1), which was discussed in Chapter 1 for simple and ideal cases.

The (x,y) color coordinates of the 15 LEDs in LED-Array-S1 were measured using both instruments. These, along with the corresponding luminance and illuminance data, are presented in Table 4.5. It was encouraging to note the (x,y) color data from the two instruments were in reasonable agreement.

The desired color temperature in white decorative lights is subjective. However, they should not be exceedingly high with high blue spectrum content because many viewers find that uncomfortable and it may have negative photobiological effects at night. The CCT data could not be obtained with known accuracy, perhaps due to the lumen output being too small from the individual LEDs in LED-Array-S1. Qualitatively speaking, the LEDs' color temperature appeared fairly cool and uniform. The (x,y) color coordinate data from CL-500A are shown in Figure 4.27.

Despite the large variations in the (x,y) coordinates and the data indicating very high CCT for the 15 LEDs seen in Figure 4.26, these types of LED arrays are often deemed acceptable for decorative lighting. Since these small, discrete lamps

Table 4.5. Measured Luminance, Illuminance, and Color Data for LED–Array–S1 Sample Using Two Different Instruments

| Data Taken: July 16, 2012 | CS-100 Data | | | Data Taken: July 17, 2012 | CL-500A Data | |
| | Luminance | Chromaticity | | Illuminance | Chromaticity | |
Sample	(cd/m²)	x	y	(lm/m²)	x	y
1	419	0.246	0.278	241.30	0.2608	0.2692
2	434	0.271	0.289	178.10	0.2586	0.2646
3	385	0.270	0.293	295.90	0.2649	0.2790
4	547	0.286	0.293	299.56	0.2645	0.2777
5	344	0.264	0.261	173.48	0.2630	0.2783
6	340	0.269	0.279	239.43	0.2646	0.2806
7	295	0.299	0.259	413.01	0.2508	0.2539
8	497	0.245	0.257	235.48	0.2635	0.2748
9	334	0.260	0.305	201.98	0.2560	0.2592
10	257	0.268	0.297	80.25	0.2667	0.2891
11	472	0.266	0.294	286.91	0.2626	0.2720
12	466	0.255	0.305	48.52	0.2582	0.2697
13	267	0.248	0.252	282.30	0.2584	0.2702
14	326	0.246	0.225	202.56	0.2691	0.2846
15	315	0.260	0.262	139.86	0.2585	0.2704

Note: The CL-500A data were taken with a computer, whereas the CS-100 data were recorded by hand.

are not utilized for illumination, their color rendering property is not very relevant. Nevertheless, it is important to note that such an ensemble of LEDs would not be desirable for constructing luminaires for most illumination applications.

4.3.2 Recommendations and Guidelines for LED Lighting Metrics

Measurements presented for the three categories of light in the previous sections demonstrate the importance of obtaining the right set of data to assess lighting performances correctly. While our visual assessment is crucial, quick and qualitative judgments are not adequate for many applications. One should not stare at illumination lamps and luminaires to make any metrics judgments of brightness, color, and other lighting properties. The illumination quality of lamps and luminaires cannot be adequately determined by viewing them because

1. The eye cannot fully determine from a quick glance whether the lamp brightness is adequate for illuminating objects and space of interest.

2. The eye cannot determine how much brighter one lamp is than another in actual quantity due to nonlinear and saturation behaviors of our vision.

3. Lamp brightness is insufficient for determining its illumination abilities, which depend on lumen distribution properties that require proper integration of flux over space.

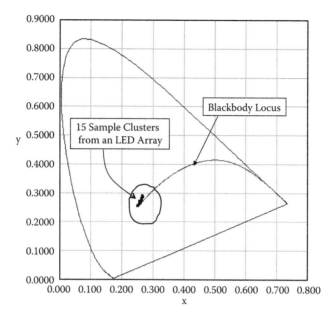

Figure 4.27. Measured color variation of 15 LEDs in LED-Array-S1 sample shown in terms of (x,y) coordinates in the CIE 1931 chromaticity diagram. All colors fall in the high CCT category with significant (x,y) coordinate variations.

Traditional lamps have undergone many design and development cycles that resulted in various types of product designs to suit many applications. LED lamps and luminaire technology have brought many more possibilities to the lighting industry. However, the unique lighting characteristics of LED emitters must be assessed accurately to utilize their full benefits in order to produce high-quality lamps and luminaires. Because LED sources can have a wider variety of spectral characteristics compared to their counterparts, the interdependencies of photometric and colorimetric properties of LEDs need to be accounted for properly in the instruments under different operating conditions. Colorimetric properties must always be measured with accurate control of either luminous flux or luminance as the reference parameter, ensuring that the operating conditions are stable. To measure color data most accurately, spectrally based instruments should be used. Traditional photometric and colorimetric instruments in the past mostly have been using the CIE Illuminant A as the reference source; further, they are RGB filter-based, where their spectral responses were configured to match closely the CIE $V(\lambda)$, $X(\lambda)$, $Y(\lambda)$, and $Z(\lambda)$ spectral functions. Such instruments are sufficient as long as the test light source spectrum nearly matches that of the reference illuminant. However, for CFLs and LEDs that may have high blue and unevenly varying spectral content, the measurement errors can increase significantly.

Color quantification of LED lamps has become a high priority in the lighting industry. Typical LED manufacturer specification sheets include color coordinates, CCT, and CRI information; some also include color purity or deviation data.

All such color properties may be calculated from the spectral data. Primarily, CRI and CCT parameters are used for matching bins. In many applications, matching LEDs to each other can provide more benefit than matching in absolute terms.

4.3.2.1 Specification Guidance for Retail Lamps

The recommendations provided in this chapter to characterize lamps comprehensively are particularly meaningful for lighting professionals. But it is impractical and ineffective for any retail lamp units to include the many parameter values described in Table 4.1 on the casing. However, for easier selection rules for the consumer, it would be helpful if primary information such as the following were provided for retail LED lamps along with an application identifier:

- Total flux

- Radiant angle

- CRI

- CCT

- Wattage

- Application (e.g., room light or desk light)

4.4 Measurement and Characterization of LED-Specific Semiconductor Lighting Properties

Thus far in this chapter, we have investigated the measurement and characterization of traditional and LED lamps. In order to optimize LED lamp performance, it is essential that various elements of LED emitter designs be verified through rigorous measurement and characterization processes during manufacturing. In Chapter 2 and Chapter 3, we have seen that LED science and technology are rather complex. Consequently, LED lighting product development requires intensive measurement and characterization techniques to verify performances expected from designs as well as to optimize the binning process. These are primarily optoelectronic, photometric, and colorimetric characterizations under different drive (including pulsed and continuous wave [CW] modes) and thermal conditions. These may be performed at the wafer, single die, or packaged stages depending on suitability or the manufacturer's production capability. Brief summaries of these characterizations are provided next.

4.4.1 Optoelectronic Measurement and Characterization of LEDs

The light output characteristics of LEDs are inherently related to the chip's semiconductor material properties. Therefore, full characterization of materials is crucial to understand and control an LED emitter's performance. These include the following:

- Electronic and other structural characterization of doped GaN and related alloys. Specifically, these include temperature-dependent Hall effect measurements that provide Hall mobility and temperature-dependent electron concentration; DLTS (deep level transient spectroscopy), which provides information on radiative recombination process efficiency; TEM (transmission electron microscopy) and SIMS (secondary ion mass spectroscopy), which identify impurities, determine doping efficiencies, and analyze surface morphology; x-ray diffraction for atomic structural information; and C–V and I–V measurements for characterizing electrical properties.

- Temperature-dependent optical characterization such as photoluminescence and electroluminescence of a GaN alloy material system under various optical and electrical excitations.

- Determination of forward voltage (V_F), total luminous flux (Φ), and dominant wavelength (at some nominal drive current, I_d).

These are some of the primary material and device characterizations for LEDs. In-depth research in these fields is ongoing within the semiconductor lighting industries and interested readers are encouraged to study these further.

4.4.2   Lighting Parameter Characterization under Varying Thermal Conditions

Photometric and colorimetric characterizations of LEDs are needed under different thermal and stress conditions to qualify them as retail lamps with specific ratings. The operating lifetime of LEDs must be determined under a variety of stress conditions, including high temperature (e.g., 85°C), high humidity (85%), and low temperature (e.g., –30°C) combinations. For LEDs, L70 and L50 lifetimes are generally defined for two different lumen maintenance levels, as discussed in Chapter 3; colorimetric lifetime should also be specified using, for example, certain acceptable shifts in the dominant wavelength and CCT. It is important that the thermal performance specifications included in the manufacturer's data sheet be detailed and reliable since most lighting designers and end users are unable to confirm thermal measurements or their accuracies. This is particularly critical for replacement products because LED lamps retrofitted into a variety of existing fixtures cannot be verified for their thermal performances very easily. For most reliable thermally dependent lighting specifications, it is best to design and specify LED luminaires for integrated systems rather than for discrete generic lamps. This is true because each integrated luminaire design bears its own thermal challenges that depend on the number of LED chips, their drive current, housing construction, and other factors.

Comprehensive thermal and lighting characterization of high-power LEDs is a unique proposition that requires specially constructed measurement systems. GL Optic (a subsidiary of Just NormLicht GmbH, Germany) is one company that

Figure 4.28. A photograph of a GL OptiSphere 1100—an integrating sphere with a 1.1 m diameter. It measures thermal transient behaviors for photometric and colorimetric parameters of various lamps including LED light sources. (Photo courtesy of GL Optic GmbH.)

offers integrating spheres capable of providing such characterization systems that combine thermal, photometric, and colorimetric testing in a stand-alone system that may be integrated and scaled within a production line with certain additions. Thermal transient measurements, in particular, can provide such critical information as internal thermal resistance for high-power, packaged LEDs, which can enhance calculation and modeling capabilities. These can also help the binning process by identifying failures. Figure 4.28 shows such an integrating sphere from GL Optic that can measure thermal transient variations for a full set of photometric and colorimetric parameters of LED engines and lamps.

For high-efficiency throughput, it is important to conduct batch characterization of LED devices in a nondestructive manner whenever possible. If the binning measurements only last a few milliseconds, the measured characteristics will be those corresponding to a much cooler junction temperature, as opposed to the actual operating temperatures of most real-life applications. Therefore, in order to obtain consistent and realistic results, LED devices under test should be burned in to a stable condition before taking photometric and colorimetric measurements at various temperatures to minimize thermal dependencies.

Because LED lamps and luminaires are manufactured using multistage development platforms, their testing and qualification should be carried out through various production phases, including:

- Wafer-level for inspection and binning to verify electrical and optoelectronic properties

- Individual LED die for testing before assembly to detect changes due to soldering and wire bonding

- Assembled single LEDs in submounts for measuring thermal performance

- Assembled LED arrays on submounts for measuring heat dissipation to the PCB

- Assembled LED arrays with secondary optics for measuring optical efficiency

Comparison of final lamp or luminaire characteristics with those from the preceding items allows manufacturers to assess the overall confidence of all measurements.

4.4.3   Standard Activities for Photometric Measurements

The LED industry has been making notable progress with establishing photometric measurement standards and guidelines in the past few years. The recent updates and releases of Illuminating Engineering Society (IES) LM-79, LM-80, and ANSI/NEMA/ANSLG C78.377 standards are such examples. The US DOE also published a preliminary document, "Manufacturer's Guide for Qualifying Solid-State Lighting Luminaires," that outlines performance benchmarks for its "energy star" criteria. The document lists DOE's certified facilities that conduct energy star qualification testing and details approval procedures. Additional such programs have begun accrediting qualifying instruments and laboratories for specific testing capabilities.

This establishment originated from a US energy policy act in the 1990s that required the DOE to specify energy-efficient standards for certain types of fluorescent and incandescent lamps by means of particular test procedures that can be conducted by accredited laboratories using applicable IES and American National Standards Institute (ANSI) standards [92]. Consequently, the Lighting Equipment Division of NEMA (National Electric Manufacturing Association) urged that NVLAP (National Voluntary Laboratory Accreditation Program) establish a program for laboratories to test lamps and luminaires of interests. NVLAP, NIST (National Institute of Standards and Technology), and DOE have more recently been building upon the path that began in the 1990s for LED lamp and luminaire testing to facilitate the development and implementation of standardized photometric testing for solid-state lighting products.

Such collaborative efforts helped realize LM-79 and LM-80 standards and their extensions and developed proficiency testing artifacts to support NVLAP accreditation processes. Their guidance enabled numerous laboratories in the

United States and abroad to receive or to be in the process of receiving NVLAP accreditation for LED testing. A few other accreditation organizations around the world are also forming to support LED lighting product development and testing to measure thermal, optical, and electrical properties of LEDs, LED arrays, and various different finished solid-state lighting (SSL) products comprehensively.

The CIE and DIN (Deutche Industry Norm) guidelines are also aiding instrumentation companies around the world to design and manufacture turnkey-type solutions for R&D and production testing for measuring radiometric, photometric, and colorimetric parameters of standard and high-power LEDs. LEDs have numerous applications in the lighting and display industries and therefore many more standards and guidelines will need to be adopted in the near future. Since lighting and display products provide distinctly different functions, primary lighting metrics for various illumination sources supporting the two categories need to be distinguished accordingly.

5 LED Lamp Design Considerations

5.1 Introduction

From the results and discussions in the previous chapters we are now familiar with the idea that to produce high-quality illumination, lamp and luminaire designs must be suited to the intended applications. Incandescent and fluorescent lamps that we see today are optimized not only based on their sizes and shapes required to address broad applicability, but also based on manufacturing constraints. Many technologies often face limitations that primarily result from inherent phenomena, which affect manufacturing feasibility and in turn compromise product performance. For traditional lighting, it has been practical for manufacturers to produce a small variety of general category lamps and have lighting professionals primarily construct luminaire designs to suit many different applications. The lighting industry thus far has been largely addressing the different illumination requirements of specific applications by designing suitable luminaires around these general category lamps; understandably, then, the field as a whole became divided into two distinct groups: lamps and luminaires.

However, for LED lamps and luminaires, the problem is now different. Because LED lighting technologies face certain manufacturing limitations of their own, a basic lamp to suit broad illumination is difficult to produce. Traditional luminaires or any of their variations placed around many different types of LED lamps produced thus far are simply ineffective in generating the desired illumination for general applications, even when an array of emitters is used in a variety of configurations. LED lighting designers face unique challenges in adopting the same principle of traditional lighting that is accustomed to utilizing incandescent and fluorescent lamps because a standard, omnidirectional or even a reasonably broad-directional LED lamp of a certain practical size remains difficult to produce. In this chapter, we first examine some general lamp requirements for various applications. Then we explore some LED lamp designs to suit a number of typical applications. In the last section of this chapter, we look at the important trade-offs that many LED lamps encounter, which are intended to serve as the bases for optimizing LED lamp designs.

5.2   Lighting Applications and Lamp Requirements

The lighting industry is currently experiencing a colossal growth and paradigm shift due to population growth, industrialization, and the introduction of new technologies. Solid-state lighting (SSL) is one such novel technology that not only offers more energy efficient solutions for many existing applications, but also is able to introduce exciting new applications that enhance people's lives. However, such different and intricate technology also brings challenges and perplexity to the lighting industry that has developed over 100 years. A majority of the people around the world have come to expect certain levels of illumination performance from artificial lighting at home and in commercial arenas. These consumers are also very familiar with conventional lamp and luminaire usage, their wide accessibility, and their inexpensive prices.

While time will unfold many different LED lighting applications and their benefits, it is now important for LED lighting professionals to recognize the current widespread general lighting solutions already adopted worldwide and carefully assess their performances and benefits. Such studies will then provide the necessary guidance for SSL developers to design and manufacture products that can maintain or exceed the lighting quality to which people have become accustomed from certain incumbents, while improving on energy efficiency with next-generation LED counterparts. Here, then, we consider the following five common lighting applications and their corresponding features or requirements needed from lamps or luminaires.

5.2.1   Ambient Lighting for Residential and Commercial Applications

General purpose ambient illumination has been the oldest and the most ubiquitous application since the dawn of artificial lighting invention. It entails providing illumination over wide spaces in home and work environments. Long fluorescent tubular lamps, also known as linear fluorescent lamps (LFLs), began replacing incandescent lamps primarily in commercial buildings after their introduction in the 1930s because of their much larger sizes and higher energy efficiency. However, even today, a large number of people still prefer incandescent over fluorescent lighting for space illumination at home and in some commercial environments, despite the availability of compact fluorescent lamps (CFLs) that fit into the Edison sockets.

As color and other properties of fluorescent and LED lamps are improving, incandescent lamps are facing real threats of becoming obsolete. In fact, several countries, including Switzerland, Australia, parts of Canada, the European Union (EU), and Great Britain, are already preventing further sales of certain wattage incandescent lamps [93–97]. In the United States, many state and federal legislators have been ambitiously pursuing enforced phase-outs of incandescent lamps starting at 100 W and ending with 40 W over the upcoming years [98,99].

Despite being very energy inefficient, incandescent bulbs generate very desirable ambient illumination in terms of color, light distribution, and brightness. Their illumination is aesthetically pleasing, which can be further enhanced with their ability to dim gradually when desired. In contrast to other lamps, they can be easily placed at any height and virtually anywhere in a room via portable luminaires without requiring immediate access to built-in electrical sockets or luminaire housings. These features make them very desirable for residential and certain commercial applications where artistry is of great importance.

Ambient lighting requirements depend on the room or space dimensions, applications, and the typical background light expected from daylight or neighboring artificial light sources. Residential applications generally include dining, watching TV, lounging, and entertaining guests—but not performing tasks such as reading, writing, drawing, or sewing, where visual acuity is important; such special needs can be met with augmented task lamps when desired. Commercial applications may include selective restaurants, hotels, and art museums when their owners tend to be devoted to illuminating their subjects beautifully and preserving color fidelity and three dimensionality that require light sources to have well-balanced color, brightness, and light distribution properties.

Incandescent lamps commonly used for ambient illumination are 40, 60, and 75 W A-line bulbs, which provide nearly omnidirectional light distribution with total flux typically ranging from 500 to 1000 lm. Compact fluorescent lamp counterparts provide 900 to several thousand lumens starting at only 14 W with a nominal efficacy of 65 lm/W. Although their spatial light distribution is not as uniformly omnidirectional as the output from incandescent lamps, luminous flux from CFLs does spread over very wide angles and illuminates much larger space volumes. Consequently, CFLs are more effective in providing the desired ambient illuminance levels over larger space, consuming much less energy. The compromise is mainly lower color quality for most applications. Many LED replacement lamps that currently provide somewhat better color quality compared to CFLs fail to generate equivalent illuminance levels over comparably wide angles covering similar spatial dimensions above and below the lamps.

While creating ambient lighting in large homes and commercial environments with only incandescent lamp technology consumes a great deal of energy, it does provide attractive illumination for certain applications. In order to craft the best visual experience for both customers and employees, certain restaurants in New York City only use incandescent lamps known as "filament" or "vintage" lamps that produce substantially less luminance and flux compared to regular incandescent lamps for the same wattage [100]. Figure 5.1 shows a photo of the Roman-style trattoria, Maialino's, in the Gramercy Park Hotel in New York, which contains vintage lamps. These illuminate the various food dishes with the best possible rendering of their wide color ranges while creating a wonderful ambience for the guests.

Figure 5.1. Maialino's, a Roman-style trattoria in the Gramercy Park Hotel in New York City, uses filament lamps as shown in this photograph. These lamps illuminate their many food dishes with ideal rendering of wide color ranges, while creating a wonderful ambience for its people. (Photo courtesy of Maialino's; photo credit: Ellen Silverman.)

Figure 5.2 shows a photo of a vintage (filament) bulb that is rated for 40 W; however, it produces only about one-third of the illuminance on planes parallel to the lamp length in any direction within several feet from the lamp, compared to that generated by a regular incandescent bulb of the same wattage. Consequently, these filament bulbs yield a much lower luminous efficacy than regular incandescent lamps.

These types of lamps are available for residential usage in the United States at such home-decor retail stores as Restoration Hardware, Pottery Barn, and Anthropologie for $9 to $20 each. When some compromises can be accepted, most ambient illumination can be created with more energy-efficient light sources, while using a few vintage lamps as accents.

Figure 5.2. A 40 W vintage filament lamp that creates a very desirable dining and gathering experience. These special incandescent lamps have a lower color temperature and luminous efficacy than regular incandescent lamps.

Traditionally, chandeliers have been utilized both as a central decorative piece and ambient luminaires to produce a reasonable level of illumination all around. This is possible in a chandelier when many bright and glowing incandescent lamps are arranged with some central point symmetry; such an arrangement can generate nearly omnidirectional and fairly homogeneous intensity distribution. Chandeliers with LED candelabra lamps lacking omnidirectional features can only serve as decorative pieces and will not be very effective for ambient illumination. Light distribution from these luminaires will not cover a very large region with good homogeneity, but nevertheless could be desirable if they are primarily used as decoration pieces.

5.2.2   Downlighting for Home and Commercial Applications

Downlighting is distinguished from most ambient lighting by its placement feature. Downlights are usually placed against the ceiling and are often used to provide partial or overall ambient lighting depending on the illumination application and space

Figure 5.3. Recessed downlights in an elevator where incandescent lamps are used. The warm CCT and high CRI of these lamps help create a pleasant visual experience for people utilizing this transportation platform in a commercial building.

dimensions. Since downlights only illuminate areas in downward directions, their lumen distribution only needs to be spread over the hemispherical region below the luminaires. Creating the necessary illumination beneath a ceiling, for example, involves controlling the light distribution within this region. Conventional downlights are luminaires that use incandescent or fluorescent lamps, either within a fully recessed structure, or in a partially protruding configuration. Figures 5.3 and 5.4 show variations where partially reflecting incandescent lamps are used to illuminate either art objects or areas where it is important to make people appear in a flattering manner.

Incandescent and fluorescent lamps produce adequate luminous intensity from a larger surface area, allowing them to spread the luminous intensity distribution (LID) from the lamp surface farther compared to current LED counterparts. Therefore, fewer incandescent and fluorescent downlights are needed to illuminate large areas. Without appropriate secondary optics, LED downlights only generate appreciable LIDs within a much narrower region. Many such luminaires, which are constructed with an array of LED engines on a flat surface, are now offered in the market, as shown in Figure 5.5. Such arrangements, when extended to larger luminaires can produce rectangular tiles, which allow broadening of light distribution over a larger region.

However, this is not effective toward homogeneously spreading the luminous intensity distribution over broader angles (i.e., light emanating from these big panel luminaires remain largely directional). Such distribution from flat panel light sources is not conducive to illuminating three-dimensional (3-D) objects and will not meet the criteria from many selective users, including artists and photographers. However, as shown in Figure 5.6, luminaires manufactured by Zenaro

Figure 5.4. The protruding recessed lights on the ceiling preferentially illumi-nate art objects. The incandescent lamps used in these luminaires help accentuate colors and shades of paintings and art objects in a welcoming manner.

Figure 5.5. LED downlight, QL-150, a commercially available 15 W downlight using 15 discrete LED modules on a flat surface. The absence of suitable sec-ondary optics will not enable *uniform* flux distribution over wide angles. (Photo courtesy of Quasar Light Co Ltd., Shenzhen, China.)

Figure 5.6. Zenaro Lighting offers recessed LED panel lights as well as other types of luminaires for illuminating office buildings as shown in this photograph. Such lighting is effective when augmented with daylight because the LED luminaires have the ability to be adjusted according to the varying sunlight. (Photo courtesy of www.zenarolighting.com)

Lighting may be effective when augmented with daylight or other light sources that provide multidirectional and nearly homogeneous lighting in office environments.

5.2.3   Large Space Lighting for Industrial Environments

Developed and developing countries use a substantial fraction of the total global energy consumption as well as a dominant fraction of total lighting energy consumption worldwide for creating illumination in *industrial environments.* Industrial applications include lighting in factories, warehouses, large retail stores, hospitals, and many others. Many of these buildings typically have very high ceilings exceeding 25 ft, with appreciably large square footage. Linear fluorescent lights (LFLs) are widely used for these applications because they can be easily manufactured to various lengths and diameters, enabling light distribution over very large spaces from extraordinarily high ceilings. The length and diameters of LFLs are chosen based on the illuminated space dimension requirements. Figure 5.7 shows two 10 ft T12s placed near the ceiling of a tall commercial building; these are still used in large factories and retail stores with ceilings as high as 25–35 ft.

Luminaires for these LFLs include ballasts to provide electrical power and housings with reflectors to direct all source light downward efficiently, spreading over a good portion of the lower hemisphere in a fairly even manner. In many factories, warehouses, and stores, a majority of the luminaires must remain at ceiling heights because tall transporting carts need to be maneuvered through the aisles that require sufficient clearance. In buildings where such provision is not necessary—for example, in gymnasiums, drop-down luminaires (as shown in Figures 5.8 and 5.9) may be used to reduce the overall flux requirements by avoiding illumination above certain heights.

Figure 5.7. Two 10 ft T12s are placed close to the ceiling in a tall commercial building where the ceiling height is over 25 ft. When used in arrays either against the ceiling or in a drop-down configuration, these can generate a vast amount of luminous flux over a very large square footage.

Figure 5.8. Drop-down LFL luminaires in a heavy traffic-oriented region of a gymnasium (Brookdale Community College, Lincroft, New Jersey). Such usage of luminaires is rather effective in providing the necessary illumination for the users without having to place the lights against the ceiling.

Figure 5.9. An exclusive area of a gymnasium (Brookdale Community College, Lincroft, New Jersey) where people tend to interact with one another more than in other parts of the arena and thus lighting of a more personal nature is appropriate. Compact fluorescent lamps with warm CCTs are used in the luminaires.

5.2.4   Outdoor Lighting for Creating Visibility

The outdoor lighting market segment makes up a significant part of the overall lighting industry that is expected to expand further in many parts of the world. These include street, parking lot, gas station, and roadway lighting, among others. Currently, high-intensity discharge (HID) lamps such as metal halide (MH) and high-pressure sodium (HPS) bulbs are widely used for outdoor lighting because of their higher efficacy over other types of traditional light sources. Many manufacturers are pursuing LED replacement lamps with the perception that they can outperform MH and HPS lamps in efficacy and CRI (color rendering index) while providing comparable flux, longer life, and much faster starts. However, as in other applications discussed previously, uniform and wider angle flux distribution providing the required surface illuminance will be the performance challenges for LED replacement lamps.

Energy efficiency and effectiveness (depends on preferential directionality) are high-priority requirements for outdoor lighting since many lamps are utilized over long hours. Other factors include durability, safety, color quality, light distribution, lumen depreciation, life span, glare, and cost. Despite a low nominal CRI of only 22, HPS lamps provide the most satisfactory average illuminance levels over broader areas per unit of energy consumed; therefore, they are still preferred for many large parking lots. Figure 5.10 shows a fairly large parking lot illuminated with pole luminaires that use HPS lamps. Where CRI is more important, MH lamps are used instead; examples include high-end mall parking lots, car dealer lots, and others.

Figure 5.10. Parking lot luminaire poles that use HPS lamps on the campus of Brookdale Community College in Lincroft, New Jersey.

Figure 5.11. An outdoor LED pole-top mounted luminaire lighting pedestrian areas on the campus of Brookdale Community College in Lincroft, New Jersey.

Because pedestrian walkways tend to have smaller square footage compared to parking lots, LED luminaires are making their entrances. Figure 5.11 shows an outdoor luminaire with nearly 50 individual LED modules arranged in a nonrectangular grid pattern to illuminate pedestrian walkways.

5.2.5   Ultrahigh-Brightness Lighting from Confined Surfaces

There are a number of applications where much higher levels of luminance are required from lamps than those of most lamp types. These high-luminance lamps are utilized for automotive headlights and various projection applications. While luminance levels of most lamps are limited to several hundred nits, automotive headlights and projection lamps require several tens of thousands of nits. Automotive headlights—or more accurately, headlamps—are luminaires that attach to a vehicle front in order to provide long-range forward visibility at night or during poor weather conditions. Car headlamps have two white light sources that provide high and low beams. High beam, also known as "main" or "full beam," generates a very bright and symmetric light distribution straight ahead to provide visibility over a very long viewing distance; however, such brightness is blinding for other drivers on the road and therefore is only appropriate in the absence of other cars. Low beam, also known as dipped or passing beam, provides sufficient forward and side illumination while limiting the amount of light and glare directed toward drivers approaching from the opposite direction. Low beams are designed to create an asymmetric light distribution to direct light preferentially downward and rightward for right-handed traffic, or leftward for left-handed traffic.

Currently, most automotive headlamps use filament (tungsten, halogen, or a combination) or HID lamps for high and low beams, and a few luxury cars use LED lamps mainly for low beams [101–103]. The 2009 Cadillac Escalade Platinum was the first vehicle that used all LED headlamps in the US market [104]. Figures 5.12 and 5.13 show LED headlamps manufactured by Osram Sylvania used in Audi A8 automobiles.

Although LED lamps can reach much higher luminance than halogen lamps, they do so only over a much smaller area. When several such high-luminance chips are placed very close together in order to produce sufficient luminous flux while maintaining the luminance level, it presents a very difficult thermal management problem. In addition, the accumulated heat in the LED modules as well as high heat from the engine compartment compounds the problem, further degrading headlamp light output performance over time. As such, the LED

Figure 5.12. The headlights in this Audi A8 use LED headlamps manufactured by Osram Sylvania. (Photo courtesy of Osram Sylvania.)

Figure 5.13. The interior circuit board of Audi A8's LED headlamps manufactured by Osram Sylvania. The rated life span for this headlamp is 7,000 hours.
(Photo courtesy of Osram Sylvania.)

headlamp life span in Figure 5.13 is only 7,000 hours, which is significantly less
than that of most LED A-line bulbs.

While LED headlamp technology improvements are being pursued, they have
already penetrated the automotive market for parking, brake, turn signal, and
daytime running lamps. These applications face significantly fewer design challenges than those encountered for headlamps and are able to utilize such benefits
as lower power consumption, durability, and packaging flexibility.

5.3 Designs to Suit Lighting Applications

Lighting designers create plans to provide lighting over desired spaces by utilizing many of the basic lamp and luminaire features described in the prior section.
They often work with architects, builders, and electricians to carry out lighting
plans that entail electrical power outlet provisions in designated locations, energy
consumption, and power budgets. As LED replacement lamps and luminaires
enter the market, lighting designers will begin investigating whether their features are comparable to those of the current standard products, or if they are nonetheless favorable to creating the appropriate lighting for a particular application.

While such practice is customary for commercial and industrial applications,
most standard homes are built with some basic electrical outlet arrangements by
today's standards and residents usually create their environmental lighting themselves. For commercial or residential consumers alike, it is unclear whether any
energy-inefficient lamp phase-outs will be successful in the long run because of
health, end-to-end savings considerations, aesthetic values, and other reasons that
might unfold over time. However, one thing is obvious: Using the most *appropriate* energy-efficient lamp for a certain application by each person or a site will

save a vast amount of energy and resources that are already becoming significantly less prevalent on an average per-capita basis. Such an adoption in turn will create a cleaner and a more sustainable environment for all. So the question is, given the various choices and circumstances, how does one learn about selecting the most appropriate lamps? Reverting a step, how do the manufacturers produce the most energy-efficient and yet well-functioning lamps for various applications at affordable costs?

In order for manufacturers to produce LED replacement lamps successfully, they must focus on three key parameters *simultaneously: color, brightness,* and *spatial flux distribution.* Finding robust engineering solutions created using appropriate design optimization processes involving these parameters will lead to high-quality and reliable LED lamps and luminaires for the five typical lighting applications discussed in the previous section. The importance of these three parameters is elaborated next.

5.3.1   Lighting Quality Factors

Judging lighting quality is complicated because it requires time and experience. It is easier to judge the quality of illuminated electronic displays and electric signs, partly because we view them directly and their functions do not include illuminating other objects. In contrast, lamps and luminaires illuminate other objects, including people, and it is best when all lit objects appear as natural as they do in daylight. Color, brightness, and spatial distribution are major aspects that contribute to lighting quality, all of which are naturally embedded in daylight quite perfectly.

5.3.1.1   Color

Color properties of light are the most widely recognized and appreciated; they can be attributed to the many color scientists that produced invaluable work over the past century. These properties are fairly well defined for most light sources. In Chapter 4, we have seen their characterization through CRI, CCT (correlated color temperature), color coordinates, and spectral power distribution. Nowadays, the ratings for CRI and CCT are provided for the majority of retail lamps.

It is duly recommended that lamps that provide high-quality light have high CRI and CCT in the range from 2700 to 5000 K depending on the application. Generally, warmer CCT (lower value) is preferred for relaxed atmosphere and cooler CCT (higher value) for office and working environments.

5.3.1.2   Brightness

Brightness is often misunderstood. Brighter is not always better for general illumination or displays and should not be treated as a figure of merit unless the application is specified. Brightness or, more technically, luminance is an inherent property of a light source. For general illumination, source luminance is never rated and almost never characterized by lighting designers. Instead, for surfaces of interests, they typically measure illuminance (in lumens per square meter or some equivalent), which can be freely described as flat surface light density. However, as LED lighting products are penetrating the market and brightness continues to be misused and misunderstood in general, it would be prudent to start rating source luminance, or at least provide a maximum rating ensuring that this parameter does not exceed certain value. In contrast to general illumination, signage and display industries specify luminance more often and illuminance less often. As discussed in the previous chapters, these quantities are interdependent.

Although viewers do not look at light sources directly, if they happen to be visible from viewers' common locations, the light source should not appear excessively bright. Too much brightness not only distorts viewers' vision, but also may harm their eyes, sometimes permanently if viewing duration is long. For example, a projector light found inside the device casing should not be directly viewed for very long. The same is true for a car headlight. Accidental viewing is usually all right because our eyes close quickly when we suddenly come across a very bright light source.

So what is excessively bright? Although not always explicit, the lighting industry has implicit standards that relate to glare, and most designers know when a light source is too bright or not bright enough. Nominal brightness requirements depend on how far the source is from typical viewers and what the typical expected ambient brightness level is. Downlights are often placed on or close to the ceiling in regular rooms that may only be approximately 10 ft away from viewers' eye plane. Since most such lamps and luminaires may be used indoors throughout the day or only outdoors during dark hours, the ambient light levels are usually in the low to medium range. Therefore, luminance or brightness levels of these luminaires should be within 200–330 nits in order to keep them in comfortable ranges for the viewers.

5.3.1.3 Distribution

Unlike color properties, spatial flux distribution (SFD) characteristics of an illuminated environment are usually not very well recognized by most people. Artists, photographers, and film makers are better at judging how light distribution in space can improve illumination. As we have seen in Chapter 4, its quantification by means of goniophotometric characterization is laborious and expensive, and most lighting designers do not perform these measurements; instead, they use known data from existing lamp catalogs when necessary. Because incandescent and fluorescent lamps have been in use for a long time, their designs are fairly well optimized and the available catalog data for light distribution usually suffice for most lighting designs. However, LED lamp technology is new and there are numerous lamp designs that lack standard configurations. In addition, very often, light distribution from LED lamps is significantly different from incumbent lamps. Thus, goniophotometric measurements are crucial for designing LED lamps for achieving the desired SFD. We shall further discuss and evaluate the importance of SFD and LID measurements in Chapters 6 and 7.

5.3.2 LED Lamp Design Considerations for Common Lighting Applications

Successful lamp designs for all five lighting applications described in Section 5.2 require achieving the necessary performances in terms of color, brightness, and spatial flux distribution. For LED lamps, color requirements can be achieved with appropriate choice of phosphors or through color mixing of multiple, nearly monochromatic LED chips. Although brightness and distribution requirements need to be met through effective optical designs tailored for each application, LED lamp designers face the same basic challenge with respect to these parameters for all five applications. This stems from the fact that today's basic LED emitter is not very well suited for either omnidirectional or even wide area illumination for two main reasons: The size of a single LED chip is too small *and* it is flat, resulting in very directional and concentrated light distribution.

Utilizing an array of discrete LED emitters to scale up to a practical size and a desired shape is still a hurdle for two reasons: (1) discrete LED arrays produce discontinuous or uneven light patterns on the lamp surface that can result in substantially nonuniform illumination over larger space dimensions, and (2) the heat generated by LED lamps and luminaires must be managed using certain thermal hardware integrated with the light source, which limits how densely LEDs can be packed inside it and whether or not they can be easily placed on multiple surfaces facing many different directions. Incandescent and fluorescent lamps emit light from continuous and nonflat surfaces, directing the flux quite evenly over much larger angles; further, they do not require additional hardware for thermal management, despite producing greater amounts of overall heat compared to LEDs, because the heat produced by these lamps can be dissipated away through convection and radiation naturally.

Specific optical design requirements are now briefly discussed for the five typical lighting applications.

5.3.2.1 Ambient Lighting Design Elements

Ideally, it would be preferred to have the LID of lamps generate ambient lighting uniformly over the entire range of solid angles from 0 to 4π(sr) with the necessary SFD strengths over designated regions. However, for practical reasons, it is difficult to manufacture such ideal lamps. Nonetheless, when ambient lamps can, at a minimum, generate LID in the approximate solid angle range of $0(sr) \leq \Omega \leq 3.67\pi$(sr), lighting designers find them very useful. This basically means that when an ambient lamp cannot emit candlepower over the entire solid angle range of a sphere, it should nevertheless spread its candlepower over most of the entire solid sphere—that is, 4π(sr)—minus some reasonably small solid angle, say, an approximate 30° cone or $\pi/3$(sr) from the bottom or top of the sphere.

Standard incandescent lamps and CFLs are quite conducive to producing such LID performance. The incandescent lamp's LID coverage occurs over a fuller range—that nearly reaches 4π(sr)—than that of the CFL because the filament can be placed sufficiently far from the electrical base that hinders light radiation. In order for LED lamps to generate flux distribution over broad solid angular ranges, some manufacturers usually place LED chips on various tilted surfaces. Figure 5.14 shows such an ambient LED lamp. Even with such configurations, LIDs of today's LED replacement lamps do not have as broad an angular coverage as those generated by the incandescent or CFL counterparts; further, their LID strengths are not as high or as uniform as those from incumbent ambient lamps.

Figure 5.14. A photograph of a 60 W LED lamp from Switch Lighting with numerous titled surfaces on which discrete HB-LED modules are placed with appreciable gaps among them. The LID properties of such an ambient lamp are still expected to lack the broad angular luminous intensity distributions typically observed in incandescent and fluorescent lamps. (Photo courtesy of Switch Lighting.)

LED lamps for ambient lighting face a graver problem when greater luminance, luminous flux, and spreading distances are demanded from them. With such extended scaling also comes greater nonuniformity and pronounced glare from most of today's LED lamps. The requirements for increases in the scaling amounts depend on the application, the size of the room (i.e., the square footage), and the ceiling height. It also depends on where people are expected to gather or where objects that require visibility are. For example, in the dining area of a restaurant or cafe, brighter illumination is primarily needed at each guest table. Hence, this requires a more selective area illumination than most general space lighting applications.

As discussed previously, most types of ambient lighting require high CRI and warm CCT because of the strict color rendering requirements for a variety of objects as well as subjects such as food, clothing, and facial complexion. Creating high-quality illumination for these is of considerable interest, particularly for selective applications. The most demanding of all such applications appears to be art museums, where artists and curators are keenly aware of the higher quality illumination that comes from incandescent lamps. These lamps not only have the ideal CRI and a desirable CCT, but also their light distribution is truly amenable for bringing out the colors and three-dimensional visual effects of beautiful paintings. This is echoed by Peter Orth of the Orth House Museum, who stated,

"Without the light *bulb* [i.e., incandescent], the painting still has its colors, but it is overshadowed by the gray tones from the rest of the room. The lights help to bring out certain elements in a painting as intended by the artist or the gallery designer" [96].

When, on September 1, 2009, the EU initiated the ban on future sales of incandescent light bulbs starting first with the 100 W lamp, artists and museum owners expressed their clear disappointment, saying that the replacement energy-saving lamps would adversely alter the way in which art is viewed and created [96]. LED lamps already challenged to produce uniform and wide angle spatial flux distribution for ambient lighting face an even greater design challenge when high CRI and warm CCT are demanded in addition.

5.3.2.2 Downlighting Design Elements

LED lamps for downlighting applications currently have several advantages over incumbent counterparts—particularly for generating relatively small area illumination from reasonably small heights. As such, they are good contenders for spot and accent lighting. In particular, smaller LED downlights are becoming popular for under-cabinet lighting. Figure 5.15 shows some LED downlights used for under-cabinet lighting in a kitchen.

LED downlights are much easier to design compared to ambient lamps, partly because they only need to emit flux over a hemispherical region covering solid angles from 0 to $2\pi(sr)$, as opposed to $4\pi(sr)$. Furthermore, downlights typically provide illumination over more confined regions than those required by ambient or other lighting applications, thus making it favorable toward LED technologies. In contrast, incandescent and fluorescent lamps need suitable luminaires to create SFD over confined regions. The DOE testing program (CALiPER, Rounds 3 through 8) has found several LED down lamps to be outperforming 45–65 W reflective incandescent lamps used in the same downlight fixtures in terms of total lumen output and efficacy, with one surpassing a typical CFL counterpart [105].

However, as the light source height is increased, LED lamp and luminaire designers face similar scaling challenges as those for ambient lighting described in the previous section: inadequate LID strength, uniformity, and angular coverage. For example, in lecture halls and classrooms with high ceilings and large square footage, LED downlights using most of today's technologies will be greatly challenged to provide the necessary illuminance levels on the many desk tops. For these applications, currently it is more appropriate and economical to use either ceiling-mounted LFLs or oversized CFLs in drop-off luminaires with reflective parabolic shades.

5.3.2.3 Large Space Lighting Design Elements

Creating task-oriented illumination in large spaces is very difficult to achieve with LED luminaire technology because the amount of scaling needed is far greater than that required for common ambient or downlighting applications. Problems encountered in increasing LID strength, uniformity, and angular coverage are

Figure 5.15. LED downlights used to provide under-cabinet lighting in a kitchen. Illumination generated by such LED lamp arrays is effective if the working distance and the lit space are both small.

now much larger since *each* luminaire needs to generate several thousand or even tens of thousands of lumens that must be distributed over solid angles as large as $2\pi(sr)$ with good uniformity. Differently sized LFLs are able to accomplish this with reflectors built in the luminaires. Figure 5.16 shows a small array of T8s used in an office building where the ceiling height is approximately 10 ft. These luminaires can be replicated as necessary to illuminate the desired square footage in general office buildings of similar ceiling heights.

For higher ceilings, T8s or T12s are used despite T5s being more energy efficient. As seen in SSL luminaires, this is also a case where energy efficiency does not scale when illumination is required over much larger volumetric space. Recall Figure 5.7 in Section 5.2.3, which shows a commercial building with a ceiling height of over 25 ft utilizing T12s. So why do not large warehouses, factories, and stores—built in the last 10 years during which T5 technology has vastly improved—switch to T5s? The answer is that luminaires using T12s produce

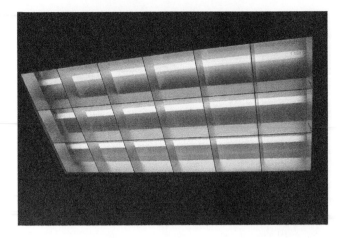

Figure 5.16. Three T8 LFLs in a luminaire shown here effectively illuminate regular office space with a ceiling height of approximately 10 ft.

considerably higher LID strength and uniformity over larger spatial ranges that are more appropriate for buildings with higher ceilings.

LED luminaires constructed with small, discrete, directional, flat light emitters *and* lacking appropriate secondary optics give rise to three main problems: (1) low LID strength, (2) LID confined to small angular ranges, and (3) pronounced luminance nonuniformity in large-array LED luminaires. If scaling is not achieved to match the LID characteristics of LFLs to address *all* these categories, LED luminaires are not likely to be successful replacement contenders.

5.3.2.4 Outdoor Lighting Design Elements

Today's LED technology is somewhat better suited for certain outdoor functions than for ambient and large-space lighting applications that were discussed earlier in this chapter. Outdoor LED luminaires are most appropriate for such applications as narrow pedestrian walkways, driveways, and garden lighting where illumination over smaller regions and smaller solid angle ranges is sufficient. For reasons discussed earlier in this chapter, LED luminaires are not yet the best choices for major outdoor lighting applications that include large parking lots, highways, and major roadways. LED outdoor lamps also face much wider temperature and humidity variations, as well as harsher weather conditions, compared to those used for indoor applications.

Outdoor area luminaire designers include several common elements that address certain functions. These consist of downward fixture efficiency and biased directionality, such as downward lumens along "street-side" and "house-side" corresponding to forward light and backlight respectively. There are also different required illuminance levels for horizontal and vertical planes in the target areas. A good process for LED lighting engineers would be to design for the required parameters using appropriate software tools and to verify with actual site measurements. This will allow translation of luminaire photometry to practice and institute useful photometric reports for outdoor area LED luminaires for the industry.

It is still an early stage for many commercial LED luminaires to satisfy the requirements of street lighting. Although some LED luminaires in the market today can fit into existing street lamp fixtures, they mostly only serve as decoration lamps. Figure 5.17 shows a photo of an old-fashioned street lamp post for which Osram offers LED luminaire modules that can easily fit into the fixture originally built for HID lamps [106]. Such a luminaire module uses a construction that has

Figure 5.17. An old-fashioned street lamp post that traditionally uses HID lamps for illumination. Today's commercially available LED refurbished lamps, such as those available from Osram, may be used for decorative purposes only. (Photo courtesy of Osram Sylvania.)

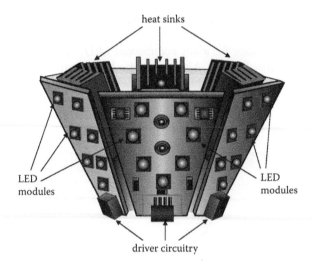

heat sinks

LED
modules

LED
modules

driver circuitry

Figure 5.18. A schematic of an LED luminaire design similar to that offered by Osram that fits into the housing of the old-fashioned street lamp fixture shown in Figure 5.17. Such a configuration can serve as a decorative lamp, but will not be able to provide street illumination like traditional HID lamps.

several large heat sinks, as shown in the schematic in Figure 5.18. Individual LED modules on flat facets in this configuration will be unable to produce the type of street illumination provided by HID lamps that have superior ambient light distribution properties. The design in Figure 5.18 can, however, serve as a decorative luminaire that would create a nice accent and ambience, allowing cost-effective refurbishing of historic street lamps.

In the last few years, many manufacturers have been touting LED replacement lamps for outdoor street and parking-lot lamps. Some evaluators have reported that in some cases, LEDs are comparable to or outperforming the incumbent outdoor street lamps [107,108]. Further careful characterization by accredited independent lighting laboratories are likely to reveal that such results do not include glare factors and that LED replacements are still underperforming HID-based street lamps in terms of illuminance uniformity. Moreover, older fluorescent technologies and designs are still being used for outdoor HPS and MH lamps, putting them at an unfair disadvantage. However, improving prior designs may require HID lamp surface sizes to increase, which may make them potentially more hazardous because of increased probability for breakage and mercury spillage. This is where LEDs have an advantage because they are rugged for outdoor use and contain no mercury.

5.3.2.5 Ultrahigh-Brightness Spot Lighting Design Elements

Designing very high-luminance LED spotlights is conceptually intriguing with respect to meeting the directionality requirements by nature. Practical lamps in this category are LED flashlights, which are popular; popularity is also increasing for projector and scuba-diving lamps. However, the problem is compounded when much higher luminance is required from significantly larger spots such as those used for automotive headlights.

Individually, LED chips are able to produce enormous levels of luminance, especially from very small areas. Generally, the smaller the area is, the higher the luminance that these single emitters can generate. Consequently, the challenge rises when designers need to scale to a very high level of luminance for larger sources. Such scaling is only possible with multiple chips placed very close together, as seen in Figure 5.13 for the Audi A8 headlamp. Full automotive headlights containing both low and high beams require a very high luminance as well as a high luminous output flux. When several high-luminance LED chips are packaged together very densely to produce the necessary luminous flux while maintaining the luminance level, the thermal management design becomes overwhelming. The current expected lifetime of such LED headlamps is only several thousand hours—not significantly higher than those for halogen lamps. Indeed, further innovation and technology improvements are still necessary for making LED headlamps prevalent in the car industry.

5.4   LED Lamp Design Parameters and Trade-offs

We have seen in the previous sections of this chapter that all five common lighting applications require certain levels of flux, flux density, or luminous intensity from larger sources covering broader angles than what current LED-based lamps can effectively offer. In general, most lighting technologies encounter trade-offs with color properties and luminous efficacy against increase in flux and flux density requirements. The trade-off between luminous efficacy and color properties is also universal in the lighting industry. Such behavior is not only apparent when one compares incandescent and fluorescent lamps, but also is verified within such incandescent lamp categories as standard, halogen, and vintage lamps. LED technologies are also no exception to many such trade-offs.

5.4.1   Trade-offs Specific to LED Lamps

For large-scale illumination applications, LEDs face additional hurdles because of their small size, flat shape, and temperature sensitivity. With improved technologies in secondary optics, appropriate LID distributions can be tailored for LED lamps to meet many large-scale applications. However, this will invariably involve more sophisticated thermal management techniques.

As seen in the simulation results in Figures 3.9 and 3.10 in Chapter 3, the T_J increase with approximately twice as many chips at twice the density is 42°C! Thankfully, with better thermal management, T_J can be brought down considerably. In the case simulated for nine LED modules in Chapter 3, the results presented in Figures 3.11(a) and (b) show that a reduction of 30°C can be achieved by increasing the fin length from 18 to 30 mm. Thus, it is clear that for high luminous flux and luminance, very large heat sinks are necessary.

The interdependencies in lighting parameters lead to the three main trade-off relationships for LED lamps:

1. High CRI in lieu of luminous efficacy

2. Warm CCT in lieu of luminous efficacy

3. High luminous flux and flux density in lieu of more sophisticated thermal management (translates to lifetime and color quality)

5.4.2   LED Trade-off Study from Transient Data

Let us now investigate some experimental data involving initial transient behavior of the samples, LED-S1 and LED-S4, introduced in Chapter 4. If most manufacturing processes are reliable for the lifetime of lamps, transient characteristics can usually provide some good indications for longer term illumination performances. In the transient experiments for our samples, the duration of the period was 30 minutes, which is a typical time during which many semiconductor devices reach steady-state conditions; the actual time depends on each lamp's thermal management effectiveness and ambient temperature. Figure 5.19 shows the calculated transient temperature rise from the initial turn-on point to a steady-state condition of the nine-module subsystem whose steady-state thermal behavior was presented in Figure 3.11(b). The transient calculation was performed using *Sauna*™.

The total luminous flux of LED-S1 and LED-S4 was measured using **GL OptiSphere 1100** (see Figure 4.28 in Chapter 4) to investigate the transient behavior over a 30-minute period. The lamps were turned on at $t = 0$. Figure 5.20 shows the plot of these data.

In Figure 5.20, one can see that as the temperature rises during the 30-minute period since the turn-on moment, the luminous flux for the two lamps decreases. The temperature rise of the samples is expected to follow a trend that is somewhat similar to the nine-module LED subsystem in Figure 5.19. This projection is especially likely for LED-S4, which we know has a lattice of emitters on a flat surface and is likely using a heat sink with fin lengths not exceeding 30 mm (see Figure 4.16). It is thus very encouraging to see that, for LED-S4, the largest percentage drop in luminous flux takes place within the first 12 minutes of the entire transient period, since the simulation in Figure 5.19 also shows that a steady-state condition is reached within the first 12 minutes!

In the case of LED-S1 (photo shown in Figure 4.15 in Chapter 4), the decrease over this period in 5.76%, where it is 8.12% for LED-S4 (photo shown in Figure 4.16 in Chapter 4). A more sophisticated heat sink in LED-S1 leads to lower flux degradation over the transient period. This comparative trend between the two samples is likely to be similar for longer duration and over their overlapped lifetimes.

The measured (x,y) color coordinate data of LED-S1 are shown in Figure 5.21 in a truncated CIE (International Commission on Illumination) chromaticity

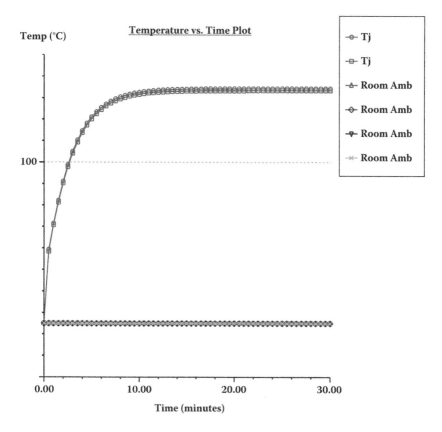

Figure 5.19. The transient calculation plotted for the nine-module subsystem (see Chapter 3, Figure 3.11b) showing temperature rise of T_J for the center module (circular symbol), for all the edge modules (square symbol), and ambient room temperature in four different locations (other symbols) around the entire subsystem remaining at 25°C. The plot shows that the steady-state condition is reached within 12 minutes. The calculations were performed using *Sauna*™ thermal software.

diagram, which is regionally magnified near the data points in order to observe the variation more effectively.

The color variation seen in Figure 5.21 for LED-S1 is not very significant during the initial 30 minutes of operation. The observed color variation for LED-S4 was less than that for LED-S1. While the first 30-minute performance appears to be good for these new lamps, much longer term uses are likely to degrade LED lamps, in general, that have poor thermal management.

Although LED-S4, lacking a more elaborate thermal management design, has shown good optical performance at room temperature, it is expected to be worse for elevated temperatures. However, usually it may not be a concern for a task

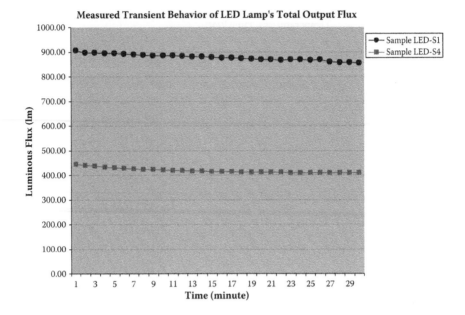

Figure 5.20. The transient flux behavior of LED-S1 and LED-S4 measured using GL Optic's integrated sphere, *GL OptiSphere 1100.* Data were obtained by GL Optic engineers.

light that is expected to be used at a desk top in controlled room temperatures for some limited hours per day.

Larger LED lamps and luminaires—particularly those to be used for outdoor lighting—require very elaborate heat sinks for thermal management. The schematic in Figure 5.18 shows rather large heat sinks for each facet of the luminaire, which is similar to those seen in an outdoor decorative LED luminaire offered by Osram [106]. Such a luminaire design reveals that spatial requirements for heat sinks, as well as driver circuitry, limit the density of individual LED modules and types. The spatial requirements for passive heat sinks, as those seen in Figure 5.18, are larger than those for active sinks that use forced air, water, or electronic cooling. In contrast, the LED lamp manufactured by Switch Lighting shown in Figure 5.14 uses passive liquid cooling, which allows the arrangement of many individual high-power modules within a fairly small circumference.

Each LED luminaire design addressing a particular application encounters unique thermal challenges based on the number of LED emitters to be used, the driver requirements, and the physical characteristics of the luminaire housing along with a variety of other factors. Designers can investigate a number of standard thermal management technologies to achieve desired LED junction temperatures, or they can utilize some suitable thermal modeling software to estimate the junction temperatures based on a parametric approach. Effective combination of experiments and simulation should ultimately provide the most successful designs.

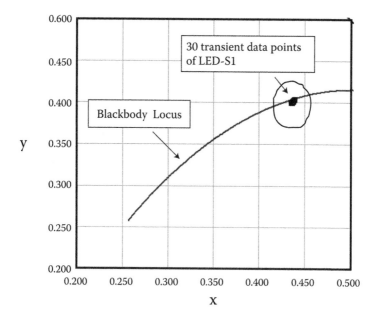

Figure 5.21. Measured transient color variation of the LED-S1 sample shown in terms of (x,y) coordinates in the CIE 1931 chromaticity diagram. Data were taken by GL Optic Engineers using a **GL OptiSphere 1100** over a 30-minute period. During this period, the LED junction temperature typically rises to a steady-state condition.

5.4.3 LED Technology Improvement Roadmaps

Current SSL lamps and luminaires all use LED arrays, which invariably produce concentrated light from small receptacles that disallows effective overlapping or aggregation of luminous flux from neighboring LED sources on surfaces and regions far away from the lamp or luminaire. Such flux aggregation, successfully produced by incandescent and fluorescent lamps, is needed to boost illuminance, particularly at large distances from the source. Most LED lamps available today lack such ability and the result is dimmer illumination at farther distances over wider regions despite their higher unit luminance and unit luminous efficacy over the incumbent counterparts. As long as the illumination plane is within a few feet from the source and only limited directional illumination is of primary interest, current LED lamps not only are satisfactory, but also generally exceed incandescent lamp performances; sometimes they are even superior to their CFL counterparts.

However, as LED lamps and luminaires are proliferating in such applications as outdoor lighting, downlighting, and general or ambient space lighting, it is important for developers to pay attention to designs that can address luminance and light distribution requirements appropriately. High-quality illumination can only be created from artificial light sources when they simultaneously include such properties as high CRI as well as appropriate CCT, luminance, and light distribution. These help viewers perceive objects and color in a natural and comfortable manner.

The LED solutions for high-quality illumination are not straightforward, partly because deviations from the current designs are likely to affect lamps' thermal management aspects adversely. The manufacturing costs can increase substantially for solutions addressing these trade-offs, at least initially. As technologies mature, costs will come down, although this is not guaranteed—and they may not come down far enough to compete with other lighting technologies for some time, especially if they also improve with better designs to suit specific applications. Nevertheless, the path to improved LED products and solutions must include determining the primary requirements for target applications, developing rigorous design and simulation tools for lamps and luminaires, building comprehensive photometric and colorimetric transient testing capabilities, and creating similar lighting industry catalogs and reports (using ".IES" formats, for example) as those provided for conventional lamps and luminaires by their manufacturers.

6 LED Lighting Design and Simulation

6.1 Introduction

In the last chapter we have seen that, in order for LED lamps to satisfy the lighting requirements of common applications, they need to provide illumination over broader angular ranges on a much larger scale compared to what the current products are able to accomplish. The current method of scaling with tiling modules is appropriate for signage and display applications because they only require incident illumination on a plane that is very near to the LED emitters. However, it is not desirable for high-quality ambient illumination, especially for those required by large-space lighting applications.

In order to develop effective design and manufacturing tools to construct LED lamps and luminaires that can scale and produce illumination over broad angular ranges in large volumetric space, an important first step would be to analyze the light pattern produced by a standard LED module. This chapter, therefore, begins with this simple numerical analysis in Section 6.2. Light distribution characteristics from multiple LED emitters are then investigated in Section 6.3 because, invariably, a single LED emitter such as an SMT (surface mount technology) module produces inadequate amounts of luminous flux for general illumination applications. In Section 6.4, we investigate some methods of scaling and distributing light over broader angles; finally, in Section 6.5, we look at some experimental results and analyze how well they agree with the conclusions drawn from the simulated cases.

6.2   Simulation of LED Light Output

In Chapters 2 and 3, we have seen some physical structures of typical high-power LEDs such as those shown in Figures 2.3, 2.4(a), and 3.2. These illustrations show a single LED chip mounted on a substrate either as a bare die or in encapsulated surface-mount modules. These fall within the basic standard LED emitters that are currently available in the market from many suppliers. The optical flux distribution of such emitters with either single or multiple dies may be simulated using various numerical methods. Ray tracing methods are most appropriate for these types of simulations because LED emitter sizes are finite as opposed to those associated with ideal light emitters such as point sources that have infinitesimal spatial dimensions or, oppositely, plane waves that have infinite physical extensions.

In addition, practical LED emitters as an ensemble often produce multiple wavelengths and have many secondary optical components as well as such light-obstructing elements as wire bonds and various types of electrical leads. A variety of these physical attributes portraying actual scenarios are best analyzed using an appropriate ray tracing method that can account for such realistic aberrations. Here we begin the simulation of a single LED emitter in a simple standard 5 mm T1-3/4 package using **Zemax,** which is a commercially available software that has recently added many appropriate real-life features mirroring those of packaged LED sources widely offered by several notable vendors in the industry [109]. Although the T1-3/4 package is not suitable for high-power LEDs for thermal reasons, their package characteristics are not consequential for simulating optical characteristics for emitters with high radiant power or luminous flux. Therefore, the following optical analysis is also relevant for other LED packages.

6.2.1   Lighting Simulation Basics of Packaged LEDs

Figure 6.1 shows the schematic of a packaged white LED placed inside a spherical detector that captures all the light rays emitted from the LED die. The results of the model are based on red, green, and blue (RGB) wavelengths, 1 million analysis rays, and 200 layout rays. The simulation was checked with 5 million analysis rays and 500 layout rays to confirm that the outputs showed good convergence even for fewer analysis and layout rays. The total LED output radiant power is taken to be 10 mW.

Let us now investigate what the detector captures. The detector output may be plotted in Cartesian or angular coordinates, which provide light distribution in position and angular space respectively. While both types of plots can be helpful for various analyses, one must be careful in treating the two differently as they have separate meanings; consequently, they have different units in their corresponding spatial or angular coordinates. Let us now look at the simulation data depicting the directivity of the source radiation in polar coordinates seen by the detector. The full directivity viewed by the detector is plotted in Figure 6.2.

The concentric circles in the polar plot provide the normalized radiant intensity ranging from 0.1 to 1.0 as shown along the vertical dotted line that separates the right semicircle from the left. The peak radiant intensity is 7.85 milliwatts/steradian (mW/sr), which occurs at ±17°. In the normalized plot shown in Figure 6.2, the value of this peak radiant intensity is equated to 1.0.

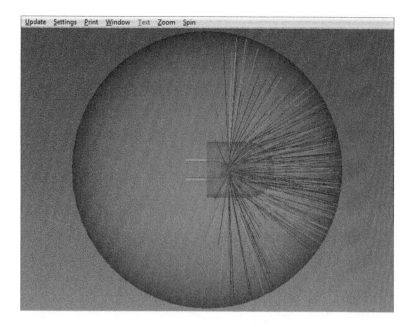

Figure 6.1. A packaged LED with a single die, electrodes, and contact wires is modeled using Zemax 12. Here, the 3-D shaded illustration schematically shows the entire LED package placed inside a spherical detector as the LED source is emitting white light composed of many individual rays, each bearing either red, green, or blue wavelengths.

6.2.2 Simulation of LED Luminous Intensity Distribution in Polar Coordinates

In addition to the directivity plot, it is also helpful to see the detector image in true colors (e.g., in gray shades) representing the luminous intensity strengths over the full angular range from 0° to 180° [2π(sr)], shown using concentric circles in a polar plot. An image of this kind is shown in Figure 6.3, where one can see that at approximate angular region of about 17° (the concentric circle of 17° radius now represents the ±17° cone in Figure 6.2), the intensity peaks just as that observed in Figure 6.2. Further, beyond the concentric circle of 30° radius, the intensity starts to diminish as seen in the previous figure, except for some very small spillage radiation seen within the angular ranges between 30° and 100° radii, after which there is virtually no radiation present beyond the 100° radius. Note that the intensity is in lumens per steradian (i.e., candelas) plotted in angular space.

6.2.3 Simulating Detector Output as Observed in Practice

It is important to note that in the example just considered, the source is placed inside a spherical detector and thus the polar plot of the detector intensity is somewhat different from that typically presented in the literature, where the detector is placed in front of the emitter and usually has a planar entrance aperture.

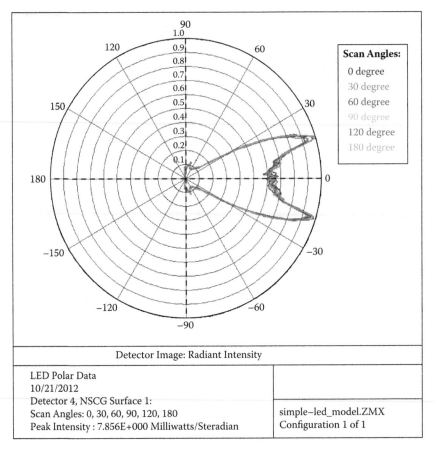

| **Scan Angles:** |
| 0 degree |
| 30 degree |
| 60 degree |
| 90 degree |
| 120 degree |
| 180 degree |

Detector Image: Radiant Intensity

LED Polar Data	
10/21/2012	
Detector 4, NSCG Surface 1:	
Scan Angles: 0, 30, 60, 90, 120, 180	simple–led_model.ZMX
Peak Intensity : 7.856E+000 Milliwatts/Steradian	Configuration 1 of 1

Figure 6.2. The full directivity polar plot of the radiant intensity distribution incident on the detector that originates from the LED emitter in Figure 6.1. The plot shows that the emitted light from the LED is very directional as the majority of the radiation is confined only within –30° to 30°. The source and the detector centers are aligned at 0°.

The latter is similar to a real-life experimental procedure used to measure LID (luminous intensity distribution) and SFD (spatial flux distribution), which was discussed in Chapter 4. Figure 6.4 shows a gray-scale polar plot of such an LID dataset measured from a high-power LED source using a detector at some distance from it. The data have been formatted into an ".IES" file. Figure 6.4 shows the LED source's normalized luminous intensity strengths in polar coordinates, which is similar to what is typically seen in the literature.

Let us now also simulate the full directivity for the same LED source whose LID data are presented in Figure 6.4. This is plotted in Figure 6.5, which is the luminous intensity distribution of the source at various angles portrayed on a plane; it shows that the LID is mostly confined within the right half of the circle, with angles ranging from –90° to 90°.

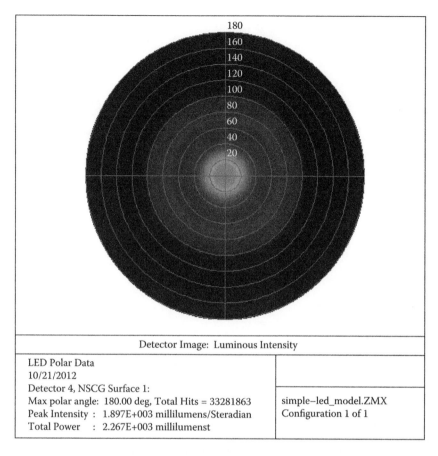

Detector Image: Luminous Intensity	
LED Polar Data 10/21/2012 Detector 4, NSCG Surface 1: Max polar angle: 180.00 deg, Total Hits = 33281863 Peak Intensity : 1.897E+003 millilumens/Steradian Total Power : 2.267E+003 millilumenst	simple–led_model.ZMX Configuration 1 of 1

Figure 6.3. Detector image representing the luminous intensity distribution generated by the LED emitter in Figure 6.1. This polar plot also shows the highly directional nature of light emitted from the LED in Figure 6.1, where the majority of the radiation is confined only within the 30° radius (i.e., –30° to 30° angular range).

Note that two different LED sources with dissimilar physical parameters are exemplified in this section. Figures 6.1 through 6.3 belong to one LED emitter; Figures 6.4 and 6.5 belong to another. For both of these LED examples considered here, the light intensity patterns seen in Figures 6.2 through 6.5 are not strictly Lambertian as expected from an ideal LED chip. This is because, unlike these examples here, the emitted light from a simple and ideal planar LED chip or die is unaltered by any wire-bond pads, wires, or certain secondary optical components that can block and smear light from different parts of the die. Further, in these simulation examples, the emitter's z-position is not set at zero, meaning the emitter itself is *not* placed in the initial x-y plane where the entire packaged LED's longitudinal position begins. (For example, in Figure 6.1, the emitter is at $z = 1.6$ mm—that is, slightly ahead from the origin of the initial plane.) Due to a

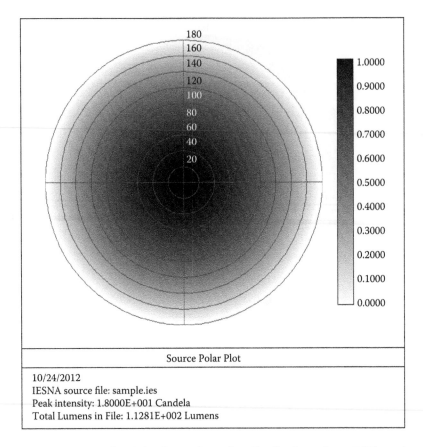

	1.0000
	0.9000
	0.8000
	0.7000
	0.6000
	0.5000
	0.4000
	0.3000
	0.2000
	0.1000
	0.0000

Source Polar Plot

10/24/2012
IESNA source file: sample.ies
Peak intensity: 1.8000E+001 Candela
Total Lumens in File: 1.1281E+002 Lumens

Figure 6.4. Normalized luminous intensity distribution of an LED emitter obtained from measured LID data, archived by IESNA in standard IES format. The peak intensity of 18 cd occurs at the center of the polar plot.

similar z-offset, the radiation pattern in Figure 6.5 shows nonzero intensity at the left hemisphere. Such a radiation pattern can be described by the formula,

$$r = a + b\cos(\theta) \tag{6.1}$$

in r, θ polar coordinates, where a is a constant representing the z-offset and b is a modulation constant for angular distribution. (The reader is encouraged to verify that Equation 6.1 is that of a Lambertian when $a = 0$ and $b = 1$.)

Today's technology allows practical high-powered packaged LEDs to produce ideal Lambertian-like light distributions when desired despite the presence of other light-blocking components. This is because small area chips are able to generate a very high amount of luminous flux and such aberrations from wire bonds and others only generate negligible variations in the output intensity pattern.

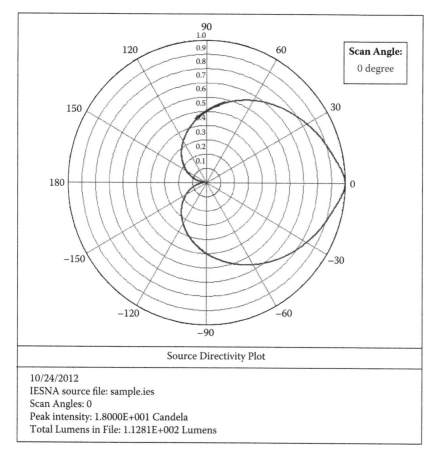

Source Directivity Plot

10/24/2012
IESNA source file: sample.ies
Scan Angles: 0
Peak intensity: 1.8000E+001 Candela
Total Lumens in File: 1.1281E+002 Lumens

Figure 6.5. Normalized full directivity polar plot of the luminous intensity distribution that is generated by the LED emitter sample whose gray-scale LID data are shown in Figure 6.4. This plot shows that the maximum peak intensity occurs at 0° as the source and the viewer's centers are aligned at 0°. While most of the radiation is found in the right hemisphere between −90° and 90°, a small amount of radiation is also seen in the left hemisphere due to the LED chip's nonzero longitudinal position.

6.2.4   Simulation of Illumination Maps for Lighting Designers

Finally, using numerical simulations, we would like to obtain some practical information or data that help lighting designers quantify some relevant parameters for light sources as well as correlate the radiant intensity data to photometric quantities that people actually visualize. A simulated planar illumination map that corresponds to the radiation pattern within practical distances is an example of such a relevant set of data. These types of data yield the SFD characteristics of a lamp at desired locations. The calculation of a planar illumination map

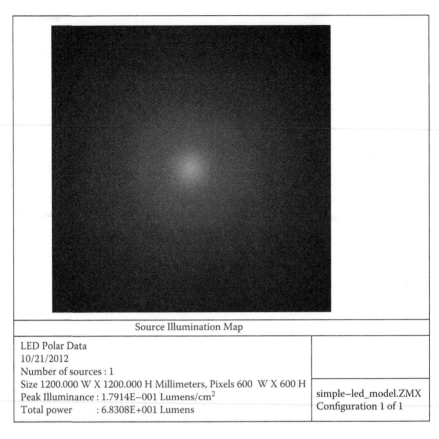

Source Illumination Map	
LED Polar Data 10/21/2012 Number of sources : 1 Size 1200.000 W X 1200.000 H Millimeters, Pixels 600 W X 600 H Peak Illuminance : 1.7914E−001 Lumens/cm² Total power : 6.8308E+001 Lumens	simple−led_model.ZMX Configuration 1 of 1

Figure 6.6. Simulated illumination map representing the luminous flux distribution on a plane 10 cm away from an LED emitter whose LID data have been measured. The photometric quantities are determined using the CIE 1931 Tristimulus XYZ color representation of the LED source spectrum.

10 cm away from the same LED source, whose measured LID data are shown in Figure 6.4, is plotted in Figure 6.6.

From the measured LID data of the light source, the illumination map data can be numerically generated at other planes of interest any distance away from the source. It is often necessary that illumination maps be simulated for finite and practical distances that may not necessarily correspond to a far-field pattern in many instances. Thus, realistic ray-tracing simulations, as opposed to simple and ideal far-field calculations, are more useful.

6.3   Scaling Light Distribution over Large Spaces

In Section 6.2 we have seen that the light distribution from a standard LED emitter is substantially confined and is primarily present over $2\pi(\mathrm{sr})$ as seen in the full directivity plot in Figures 6.2 and 6.5. The examples in Section 6.2

also demonstrate that a single LED emitter only has limited total optical power that is typically on the order of 100 lm even for high-power LEDs. Therefore, many LED emitters are necessary to generate appreciable luminous flux density over large illumination planes within practical distances from the emitter ensemble.

While it is clear that multiple LED emitters are required to scale up light distribution for illumination applications, it is usually not clear that such straightforward scaling as tiling them in a discrete fashion on a plane is sufficient to generate high-quality illumination for general lighting applications. Therefore, it is worth investigating the nature of the light distribution generated from a planar array containing several LED emitters. In order to gain a basic understanding, we begin this analysis by comparing the simulated outputs from a single high-power LED with that of two such LEDs closely spaced on a source plane. All simulations presented in this section are obtained using **Zemax 12**.

6.3.1   Single- and Two-LED System Simulations

Figures 6.7(a) and 6.7(b) schematically show one and two *typical* high-power LED sources, respectively, emitting light rays that are being detected by a 30 mm × 30 mm rectangular detector 5 mm away. The emitted power from a single LED source is taken to be 100 lm. The numerical analysis here utilizes the measured LID data from a current *typical* high-power LED source found in the IES (Illuminating Engineering Society) library.

There are several variations of optical outputs one can analyze from the configurations shown in Figures 6.7(a) and 6.7(b). We have already observed luminous

 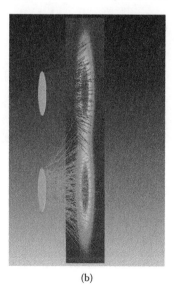

(a)   (b)

Figure 6.7. The 3-D schematics showing light rays detected by a 30 mm × 30 mm rectangular detector 5 mm away from (a) a single high-power LED source; (b) two high-power LED sources. The power from each LED emitter is 100 lm.

intensity distribution outputs in angular space in the previous section. Let us now view the optical outputs in position space at the detector plane and analyze their spatial flux distribution, which we can then relate to illumination perceived by the average human eye. Optical power distribution at the detector plane may be observed as incoherent illuminance or luminance. The latter does not imply that the source luminance properties are changed at the detector position; rather, it may be viewed as the luminance distribution produced at the detector plane from the sources. If the detector could become a light source by being able to transmit the light that is incident on it, this luminance distribution would then be its inherent property. Bearing that in mind, we plot such luminance outputs at the detector plane in position space for the two cases in Figure 6.7 and in Figures 6.8(a) and 6.8(b) respectively.

At distance of only 5 mm, the radiation from the two LEDs essentially remains just in front of the individual LEDs, forming a pattern composed of almost two

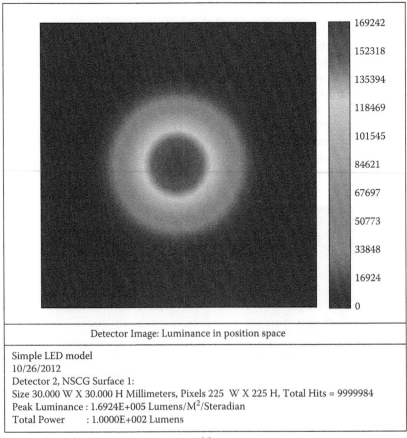

Detector Image: Luminance in position space

Simple LED model
10/26/2012
Detector 2, NSCG Surface 1:
Size 30.000 W X 30.000 H Millimeters, Pixels 225 W X 225 H, Total Hits = 9999984
Peak Luminance : 1.6924E+005 Lumens/M^2/Steradian
Total Power   : 1.0000E+002 Lumens

(a)

Figure 6.8. (a) Calculated luminance distribution in position space observed at the detector that is placed 5 mm from a single LED as shown in Figure 6.7(a). The detector detects virtually 100% of the luminous flux emitted by the LED.

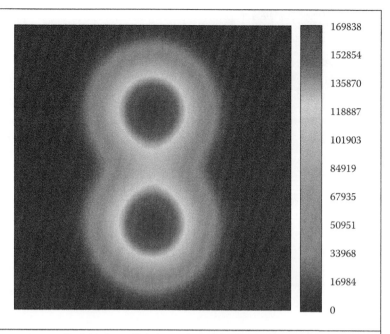

	169838
	152854
	135870
	118887
	101903
	84919
	67935
	50951
	33968
	16984
	0

Detector Image: Luminance in position space

Simple LED model
10/26/2012
Detector 3, NSCG Surface 1:
Size 30.000 W X 30.000 H Millimeters, Pixels 225 W X 225 H, Total Hits = 19927629
Peak Luminance : 1.6984E+005 Lumens/M^2/Steradian
Total Power   : 1.9928E+002 Lumens

(b)

Figure 6.8. (continued) (b) Calculated luminance distribution in position space observed at the detector that is placed 5 mm from the two LEDs as shown in Figure 6.7(b). The detector detects 99.5% of the luminous flux emitted by the two-LED ensemble.

distinct Lambertian outputs in proximity. The 30 mm × 30 mm detector detects 99.5% of the total light emitted by both LED sources. The peak luminance, occurring at the centers of the Lambertian patterns, is over 169,000 nits! Note that false color representation (converted to grayscale for black/white figures) is used to plot the luminance distribution in these figures.

6.3.2   Two-LED System Simulations at Various Distances

We now observe the radiation from the two-LED source configuration at 10, 15, and 20 mm away in order to get an understanding of the flux integration from the two LED emitters. For these distances, Figures 6.9, 6.10, and 6.11, respectively, show the schematics and their corresponding luminance distribution images in

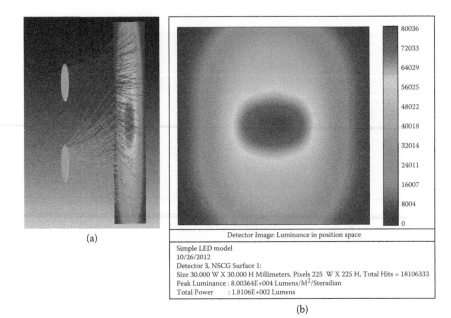

(a)

(b)

Figure 6.9. (a) The 3-D schematic showing light rays detected by a 30 mm × 30 mm rectangular detector 10 mm away from the two LED sources; (b) the calculated luminance distribution in position space observed at the detector from the two LEDs in Figure 6.9(a). The detector detects 90.5% of the luminous flux emitted by the two-LED ensemble in this case.

X-Y coordinates seen by the detector. Because the detector aperture size remains the same in all of these cases, the source images at the detector are more truncated for longer distances. Nevertheless, the peak luminance is captured in all the images while only the peripheral evanescent fields are lost.

Figures 6.9 through 6.11 show that even as the detector distance increases, the luminous flux still aggregates at the center of the detector plane, creating a concentrated high luminance area in the middle of the detector. Later in this chapter, we shall elaborate on such SFD characteristics of discrete LEDs on a plane.

6.3.3 Luminance Analysis with Varying Detector Size

The simulations presented in Figures 6.8 through 6.11 show that the two-LED configuration produces high-brightness areas over fairly small regions at the center of the viewing plane with optical power tapering off very quickly beyond the central areas. When highly intense light sources such as the two high-power LEDs in this system produce concentrated illumination only along the optical axis, a viewer looking directly along the optical axis within certain viewing distances will observe a high degree of glare. However, within such viewing distances, illumination over any reasonable size plane will be uneven and poor; significant light flux will only be present at the center of the plane over a fairly small radius within reasonable viewing distances. We verify this by increasing the size of the detector

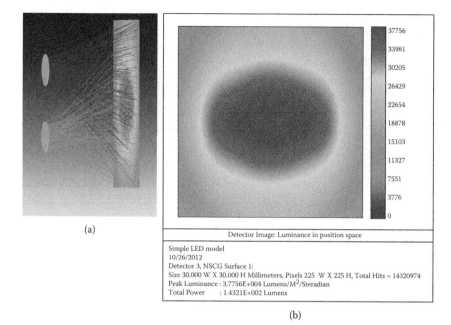

(a)

(b)

Figure 6.10. (a) The 3-D schematic showing light rays detected by a 30 mm × 30 mm rectangular detector 15 mm away from the two LED sources; (b) the calculated luminance distribution in position space observed at the detector from the two LEDs in Figure 6.10(a). The detector detects 71.6% of the luminous flux emitted by the two-LED ensemble in this case.

to 40 mm × 40 mm and simulating the luminance output at the detector 20 mm away from the two-LED system. The radiation detected by a larger detector (40 mm × 40 mm) 20 mm away from the sources is plotted in Figure 6.12 in terms of luminance distribution in position space. Although the larger detector in this case captures more power, the luminance distribution over the common position space is identical for Figure 6.11(b) and Figure 6.12(b).

It is often helpful to visualize the luminance profile in one-dimensional (1-D) plots along the X and Y center axes of the detector. We present these in Figures 6.13 and 6.14 for the two-LED system schematically shown in Figure 6.12(a).

Figures 6.12 through 6.14 confirm that although the larger detector captures more power, it captures the same maximum luminance and luminance distribution over the identical elliptical x- and y-radii as does the smaller detector over the common area; the larger detector actually shows that luminance continues to drop off sharply and monotonically toward the detector edges.

6.3.4   Simulation Output Choices: Luminance or Incoherent Illumination

At this juncture, it is a good exercise to observe the incoherent illuminance output of the two-LED systems, which is analogous to the output depicted in

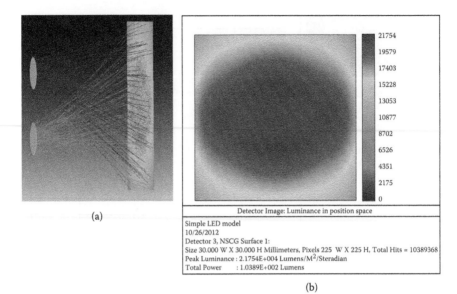

(a)   (b)

Figure 6.11. (a) The 3-D schematic showing light rays detected by a 30 mm ×
30 mm rectangular detector 20 mm away from the two LED sources; (b) the
calculated luminance distribution in position space observed at the detector from
the two LEDs in Figure 6.11(a). The detector detects 51.5% of the luminous flux
emitted by the two-LED ensemble in this case.

Figure 6.8(b). This is plotted in Figure 6.15, which shows that the radiation pattern plotted in terms of incoherent illuminance is the same as that in Figure 6.8(b),
which portrayed luminance in position space. Later in this chapter, we shall choose
incoherent illuminance rather than luminance as the parameter to describe light
distribution in position space.

6.3.5   Summarizing the Nature of LED Illumination

We are now ready to begin a qualitative discussion about the nature of LED illumination that one can deduce from the results presented here thus far. Although
the analysis presented here uses a two-LED system, it can nevertheless be straightforwardly extended to a larger two-dimensional (2-D) LED array (e.g., a 10 × 10
LED array with 100 emitters) in a source plane to assert the following:

1. Light distribution produced by standard LEDs placed on an *XY*-plane,
   with *x*- and *y*-spacing between the LEDs on the order of the emitter's
   surface dimensions (e.g., radius for circular or width and length for rectangular surfaces), will show high luminance areas on viewing planes that
   are parallel to the source plane and are within short working distances
   from the source plane. Peak and high luminance levels will be concentrated in the middle of the viewing plane because discrete LEDs on the
   source plane will all aggregate their contribution to the center of the

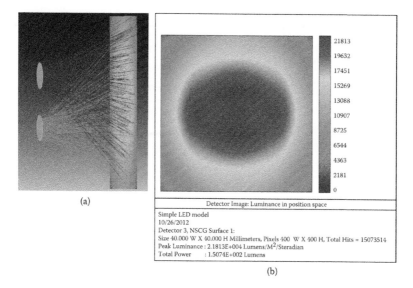

(a)

(b)

Figure 6.12. (a) The 3-D schematic showing light rays detected by a 40 mm × 40 mm rectangular detector 20 mm away from the two LED sources; (b) the calculated luminance distribution in position space observed at the detector from the two LEDs in Figure 6.11(a). The detector detects 75.3% of the luminous flux emitted by the two-LED ensemble in this case.

> viewing plane by symmetry (see Figures 6.13 and 6.14). Here we assume that the centers of the source and such viewing planes are aligned along their common optical axis—namely, the **Z**-axis.

2. From the simulated outputs, it can be affirmed that a good degree of uniformity in luminance outputs can be achieved within reasonably close distances from a general source array consisting of more than two LEDs, provided the illumination plane dimensions are close to that of the source plane containing the LED arrays. At some optimal x- and y-spacing between LEDs and at some z-position for the viewing plane, the illumination on the viewing plane can be made very uniform. That is why laptop, TV, and other display screens are effectively illuminated with LED backlighting, producing far higher screen luminance than most other technologies [110].

3. Illumination at any other plane that is not orthogonal to the optical axis will see significantly less radiation. Consequently, lamps designed with LEDs on a planar surface cannot produce good ambient illumination, which requires gradual tapering of luminous flux distribution at off-axes.

4. Luminance levels at the detector generated from a planar LED source are blindingly high as seen in the values shown in Figures 6.8(a) and (b), which are still lower than the inherent luminance level of the high-power

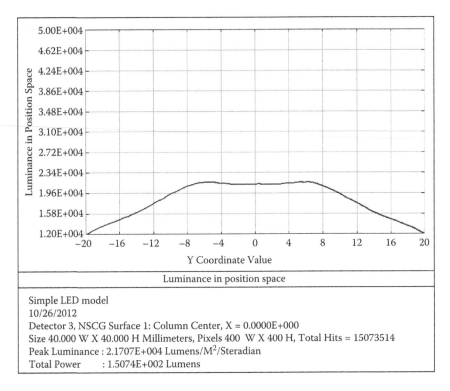

Figure 6.13. Calculated luminance distribution in position space observed at the detector along the *Y*-axis (at *x* = 0) from the two LEDs in Figure 6.11(a). The maximum luminance value stays fairly constant over only an 8 mm *y*-radius, after which the luminance drops off sharply.

LEDs considered in this analysis. In the subsequent figures where the detector is farther away, the luminance still remains very high even with contributions from only two LEDs. When more LEDs, higher luminance LEDs, or their combinations are used in a 2-D array to increase illumination at longer distances, the on-axis luminance also increases, resulting in pronounced glare for viewers directly looking at such LED light sources.

5. Light sources constructed from a 2-D LED array may provide adequate illumination for *planar* surfaces over reasonably close distances from the source, although viewers should avoid directly looking at the LED emitters. However, long-range illumination, especially over large volumetric space that includes broad angular ranges, is much more difficult to achieve with LED emitters placed on a planar surface because of the forward flux aggregation nature of discrete high-brightness (HB)-LEDs on a plane.

The preceding assertions enlighten the need for more effective light distribution scaling beyond that achieved from LED lamps that are simply constructed

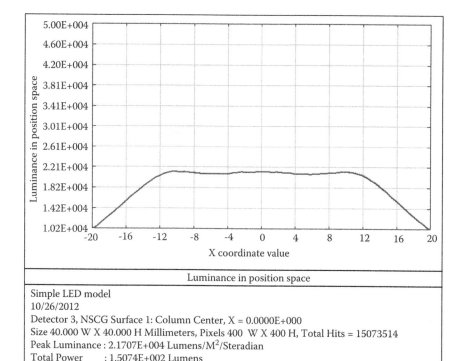

Figure 6.14. Calculated luminance distribution in position space observed at the detector along the *X*-axis (at *y* = 0) from the two LEDs in Figure 6.11(a). The maximum luminance value stays fairly constant over only an 11 mm *x*-radius, after which the luminance drops off sharply.

using planar emitters. In order for LED lamps to become ubiquitous, their design must be extended to provide adequate and desired illumination in planar *as well as* volumetric spaces of various sizes; such LED lamps should improve illumination quality by ensuring uniformity, reducing glare, and achieving farther and broader angular light distribution with fewer lamps.

The directivity simulation results presented earlier in the chapter tell us that LEDs provide substantially less luminous intensity (candela) for angles beyond 30° (see Figures 6.2 and 6.5). Certain lamps in the current market attempt to overcome this limitation by placing discrete SMT-type LED modules on tilted planes or curved surfaces. Such solutions invariably encounter numerous challenges, including more complex thermal designs and manufacturing processes. Most importantly, they still do not solve the outstanding issues concerning glare and nonuniform illumination.

The luminaire examples discussed in Chapter 5 and the numerical analysis presented in this chapter both affirm that LED lamps and luminaires constructed with small, discrete, and flat light emitters void of appropriate secondary optics are problematic for two reasons: (1) low LID strength at broader angles and (2) LID remains confined only to fairly small regions in position space, which

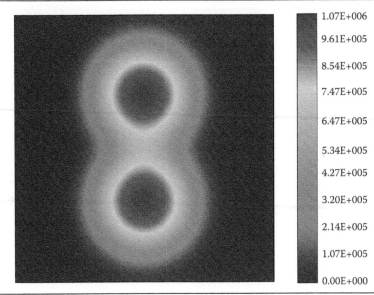

Detector Image: Incoherent Illuminance	
Simple LED model	
10/28/2012	
Detector 3, NSCG Surface 1:	
Size 30.000 W X 30.000 H Millimeters, Pixels 400 W X 400 H, Total Hits = 19927629	
Peak Illuminance : 1.0677E+006 Lumens/M^2	
Total Power : 1.9928E+002 Lumens	

Figure 6.15. Calculated incoherent illuminance distribution in position space observed at the detector that is placed 5 mm from the two LEDs as shown in Figure 6.7(b). The detector detects 99.5% of the luminous flux emitted by the two-LED ensemble and shows the same radiation pattern as that seen in Figure 6.8(b).

is a result of the emitter areas being very small. In order for LED luminaires to become successful replacement contenders for the present incandescent and fluorescent lamps for general lighting applications, their designs must address the scaling of LID required to match the levels produced by their counterparts. Hence, we now discuss some methods of achieving LED lamp designs that can appropriately scale by producing continuous light radiation from suitable nonplanar surfaces utilizing only low inherent source luminance.

6.4 Generating Uniform-, Multi-, and Omnidirectional LED Light Distribution

We have discussed much about LEDs in their current form not being able to produce as uniform and omnidirectional light distributions as incandescent and fluorescent lamps inevitably can. But what enables the incumbent lamps to produce uniform and omnidirectional radiation? And why do these lamps not generate much glare while

LEDs tend to create excessive glare for the same amount of total emitted luminous flux? The answers are embedded in the basic nature of electromagnetic radiation.

6.4.1   Application of Electromagnetic Theory to LED and Standard Lamps

Light radiation by nature follows the laws of electromagnetic waves, which are governed by Maxwell's equations. The first of the four Maxwell's equations describes how the electric field radiates from a source enclosed in a volume containing a certain amount of electric charge. This is also known as "Gauss's law," which is essentially the same as the *divergence theorem* in mathematics [111], which is written as

$$\iint_S \vec{F} \cdot \vec{n} \, dS = \iiint_V \vec{\nabla} \cdot \vec{F} \, dV \tag{6.2}$$

where $\vec{F}$ is a vector field or flux that behaves like an analytical function, meaning that it is twice differentiable. In electromagnetics, when $\vec{F}$ represents the electric field, the right side of Equation (6.2) gives the total amount of electric charge, which is proportional to the energy contained in volume V that is enclosed by the surface S. This is precisely what Gauss's law states!

Note: In the case of a light source, $\vec{F}$ may represent, for example, the position function describing the light power or luminous flux resulting from the light energy contained within the light source of volume V.

Electromagnetic wave propagation behavior falls out of Maxwell's equations in a form that is described by its electric field, $\vec{E}$, written as [112]

$$\vec{E}(\vec{k}, \vec{r}) = E_0 \, e^{-i\omega t} \, e^{i(\vec{k} \bullet \vec{r})} \tag{6.3}$$

where E_0 is the amplitude of $\vec{E}$, ω is the angular frequency, $\vec{r}$ is the position vector, t is time, and $\vec{k}$ is the wave vector given by

$$\vec{k} = \frac{2\pi}{\lambda} \vec{n},$$

where $\vec{n}$ is the complex index of refraction. The generalized index of refraction is a vector quantity whose magnitude is determined from the complex number, n_c, which can be written as

$$\left| \vec{n} \right| = n_c = n_r + i \, n_i \tag{6.4}$$

where n_r is the optical index of refraction and n_i is related to the optical absorption properties of the material.

The luminous energy that generates flux from a light source of certain shape and other physical properties, including optical index of refraction, will be emitted as radiation from the light source according to Equation (6.2) and Equation (6.3). This follows because the radiation from a monochromatic light source (i.e., an optical source containing a single frequency in the visible spectrum) is that of its electric field, $\bar{E}$, which follows the same principle as the *divergence theorem;* since any white light source is polychromatic containing a number of monochromatic lightwaves, it too radiates in the same manner via its constitutes. It then follows that any light obeying the *divergence theorem* can only exit from an enclosed light source in the directions perpendicular to the source surfaces while its monochromatic-wave constitutes satisfy Equation (6.3) as they propagate.

Therefore, a very thin LED emitter that only has one planar emitting surface will produce light primarily perpendicular to that surface. (Here we have assumed that most of the light generated inside the LED chip hit the exit surface orthonormally because the luminous flux distribution inside it is confined in a very thin sheet of active region, and any refractive index mismatch between the LED chip and the outer medium only causes a reflection loss at the exit surface.) In contrast, other light sources that have curved surfaces will produce light in all various directions normal to their surface curvatures according to the *divergence theorem.*

6.4.2   Illumination Comparison of LEDs and Curved-Surface Lamps

Utilizing the *divergence theorem,* let us now investigate the illumination behaviors of LED and curved-surface lamps, using the illustrations in Figure 6.16 and Figure 6.17 respectively.

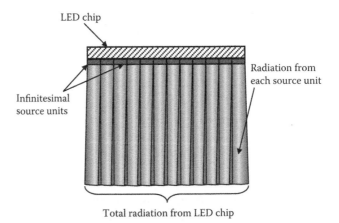

Total radiation from LED chip

Figure 6.16. Light radiation is schematically shown from a side or a cross-section of an LED chip. Not including the edge effects, the radiation is only emitted normal to the LED chip surface as governed by the *divergence theorem* or *Gauss's law.* The thickness of the source units is very small compared to the exit surface dimensions. The schematic is not drawn to scale.

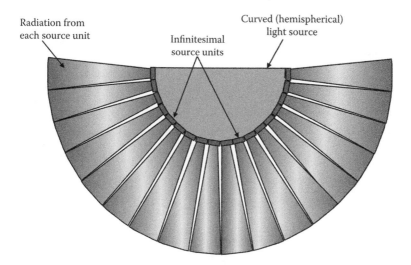

Radiation from
each source unit

Infinitesimal
source units

Curved (hemispherical)
light source

Figure 6.17. Light radiation is schematically shown from a cross-section of a hemispherical light source. The radiation is only emitted normal to the curved surface as governed by the *divergence theorem* or *Gauss's law*. The schematic is not drawn to scale.

Figure 6.16 illustrates why LEDs produce directional output resembling a Lambertian immediately in front of the source. This illustration also makes it clear that as long as LED lamps and luminaires are produced using the basic structure shown in Figure 6.16 *in a planar array,* the generated illumination will not be equivalent to that from incandescent or fluorescent lamps. The directivity of such an LED source will primarily remain over limited angular ranges and will not come close to matching those from curved-light sources.

The single LED light source schematic in Figure 6.16 may also be used to explain why LEDs produce very high luminance and sources with curved surfaces tend to produce much less of it. This is demonstrated using Figures 6.18(a) and 6.18(b).

Consider three neighboring points A, B, and C on an LED chip emitter as shown in Figure 6.18(a). First, assume that these points have some finite surface area, dS, each producing a *small but finite* cone of light as shown in Figure 6.18(a). In contrast to the analogous points D, E, and F on a curved surface for a light source shown in Figure 6.18(b), the cones of light produced by A, B, and C have much larger overlap as they all radiate in the same orthonormal direction. This invariably increases the luminous flux concentration or luminance for the light source in Figure 6.18(a). However, points D, E, and F have different orthonormal directions because they are on different curvature locations on the light source surface; therefore, the light cone overlap from these neighboring points is significantly smaller.

If we now let dS approach zero and integrate all small light cones from all infinitesimal surface points, the resulting luminance for the planar-surface LED source in Figure 6.18(a) will aggregate to a much higher value compared to that

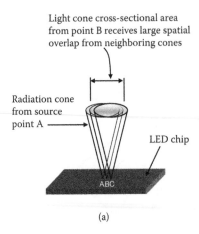

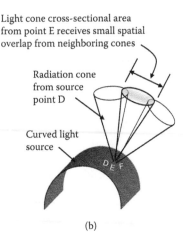

(a)   (b)

Figure 6.18. (a) Neighboring points A, B, and C on an LED chip producing light cones in the same orthonormal directions. Very high luminance levels are produced because the neighboring light cones representing luminous intensities have large spatial overlaps as shown here, resulting in highly concentrated light within small angular ranges; (b) neighboring points D, E, and F on a curved-surface light source produce light cones in different orthonormal directions and thus spread the source LID over broader angles. Significantly less luminance is produced in curved-surface light sources compared to that of part (a) because the neighboring points produce very small, spatially overlapped light cones.

from Figure 6.18(b). (The underlying assumptions for this comparison include equivalent spatial dimensions and unit luminous intensity for the two cases.) *Thus, luminance levels from LEDs, as seen in the simulations in the previous sections, are exceedingly high and invariably cause glare for the viewers.*

The preceding discussion helps clarify why high concentrations of light from discrete LED sources on a plane have the following drawback in general lighting: Illuminance levels produced by such sources at large distances are not very high because lamp surfaces remain partially dark *and* the distant locations only receive small angular LID contributions from individual and other similar neighboring light sources. If one integrates the flux density at various distances from an array of planar LED sources, the integration resulting from each Lambertian distribution produced by an LED will show that adequate illuminance can *only* be achieved for modest distances from the source and *only* over limited areas directly below the light source.

In contrast, for lamps that have continuous and homogeneous curved emitting surfaces, the distant locations receive substantial LID contributions over broad angles from individual and similar nearby lamps; thus, illuminance is boosted at longer distances through effective flux integration without generating glare. These lamp types are also well suited for generating *natural shadows* when they illuminate curved objects, whereas LED lamps of the types discussed in the analyses thus far in this chapter are not. Edison's incandescent bulbs and CFLs fall in this category; while incandescent bulbs generate better natural shadows, CFLs usually provide higher illuminance over larger volumetric spaces for the same electrical input energy.

6.4.3 Methods of Generating Multidirectional and Diffused LED Distribution

Since LEDs fabricated using the current methods fall into the light source category described in Figure 6.16, how may one spread the light concentrated in such a small spot over a large volumetric space—diffusing its lumen distribution uniformly? We now discuss several methods of creating diffused light distribution to provide multidirectional illumination suitable for ambient lighting.

Understandably, diffusion of light from LED emitters has been a primary industry interest for quite some time. Hence, many methods can be found in the literature that allow broadening of LED light outputs. Among them, the following two have received notable attention in recent times:

- *Diffractive optics*

- *Free-form optics*

Although these two methods may have some overlap, *diffractive optics* traditionally uses well-defined discrete components, whereas *free-form optics* includes the usage of a variety of custom-molded elements in a more analog and integrated fashion for controlling light. The field of optics has been engaging in *diffractive optics* techniques to diffuse and control spatial light distribution for a long time. As such, we refer readers to works by others [113–115]. Interested readers are also encouraged to learn about the utilization of *free-form optics* methods by several groups [116–118].

Here we consider a different approach, which utilizes light pipes and waveguides to broaden light from discrete LED emitters. Although certain light pipe structures may partly overlap with free-form optical elements, the overall concept introduced here to construct LED lamps for general illumination is different. Nevertheless, the approach described next could also, in part, incorporate some usage of diffractive optical elements.

6.4.3.1 Uniform and Broadened LED Lamp Output Using Light Pipes and Compound Optical Elements

We introduce a design concept to achieve uniform and broadened light from an LED lamp by using an array of discrete LED emitters, each followed by some appropriate *beam expanders* that terminate seamlessly at some curved surface from which the lamp emits its light to the environment [119,120]. In this section, we consider light pipes and compound optical elements that provide the beam expansion. Seamless termination at the lamp's outer curved surface, with respect

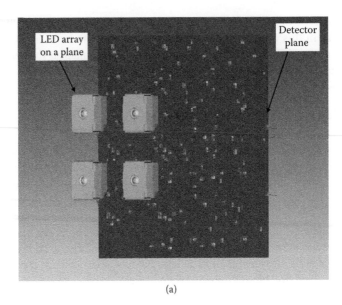

(a)

Figure 6.19. (a) The 3-D schematic showing light rays detected by a 20 mm × 20 mm rectangular detector 10 mm away from the four LED sources. The traced rays show that the majority of the light accumulates at the center of the detector in a nonuniform or nonhomogeneous fashion.

to optical and mechanical requirement aspects, may be achieved by means of some curved diffractive optical elements in front of the light pipes, ensuring that the center optical axis of each light pipe maintains its continuity and penetrates the lamp's curved surface orthonormally. In theory, these discrete elements could be integrated with the light pipes using molding techniques.

In order to see how light pipes create the broadening effect in a homogeneous manner, let us simulate the output of four LEDs with and without light pipes. These are shown in Figures 6.19 through 6.21. Figure 6.19(a) shows four packaged LEDs on a plane, and their output observed by a planar detector 10 mm away from the LED plane is shown in Figure 6.19(b).

Let us now look at the output at a larger distance from the LED plane. As previously, Figure 6.20(a) shows the four LEDs on a plane and the dectector's position; their simulated output by a planar detector 43 mm away from the LED plane is shown in Figure 6.20(b). The output in Figure 6.20(b) shows that the light at the detector is very nonuniform, scattered, and that less than 10% of the luminous flux emitted by the four-LED ensemble is detected.

Now we simulate the light output from the same four LEDs as shown in Figure 6.20(a) after they propagate through four light pipes as shown in Figure 6.21(a). The output observed by the same-size detector as in the preceding figures is shown in Figure 6.21(b).

Although the light pipes in Figure 6.21 are not optimized to capture all the light from the LEDs in this analysis, it is clear from Figure 6.21(b) that they

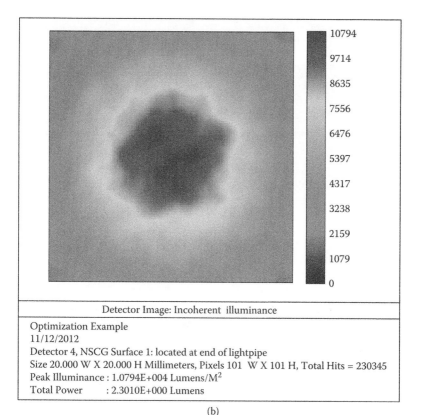

Detector Image: Incoherent illuminance
Optimization Example 11/12/2012 Detector 4, NSCG Surface 1: located at end of lightpipe Size 20.000 W X 20.000 H Millimeters, Pixels 101 W X 101 H, Total Hits = 230345 Peak Illuminance : 1.0794E+004 Lumens/M^2 Total Power   : 2.3010E+000 Lumens

(b)

Figure 6.19. (continued) (b) Calculated incoherent illuminance distribution in position space observed at the detector that is 10 mm away from the four LEDs shown in Figure 6.19(a). The output shows that the light at the detector plane is nonuniform, highly intense at the center, and that 57% of the luminous flux emitted by the four-LED ensemble is detected.

provide beam expansion in a confined and homogeneous manner. Without such secondary optics, the arrayed LEDs undergo only free-space radiation creating nonuniform illumination in space. This is illustrated in Figure 6.22.

Light pipes and other compound optical elements can prevent ineffective free-space radiation and direct light from LEDs to desired locations. Let us consider another example of such secondary optics that can broaden light in a confined manner. In Figure 6.23, we present a simulation of light broadening on a plane from the four LEDs under discussion, using rectangular compound parabolic concentrators. Figures 6.23(a) and 6.23(b) show that LED outputs can be broadened to a larger surface far away from the source without incurring much optical loss through arbitrary radiation.

Although the broadening effect shown in Figure 6.23(b) is suitable to provide illumination over a planar surface with good uniformity, it is not favorable for

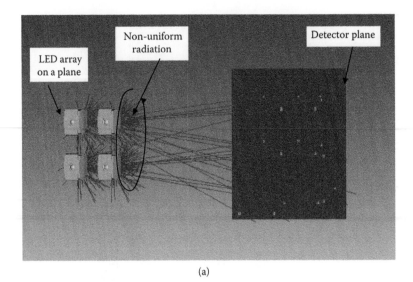

(a)

Figure 6.20. (a) The 3-D schematic showing light rays detected by a 20 mm × 20 mm rectangular detector 43 mm away from the four LED sources. The traced rays show that only some scattered light reaches the detector plane and the radiation just in front of the LEDs is very nonuniform.

uniformly illuminating volumetric spaces. To achieve such illumination, an optical design using light pipes (e.g., cylindrical) of different tapers and bends may be utilized to reach the lamp's outer curved surface. This concept is schematically drawn in Figure 6.24.

As illustrated in Figure 6.24, light pipes may be used to propagate light from each LED in a confined manner in order for it to reach the curved surface of the lamp while satisfying the continuity and orthogonality requirements of the optical axis. This will allow light to be radiated from the lamp's surface uniformly in multiple directions. To achieve adiabatic (i.e., slowly varying) and seamless optical and mechanical performances needed for such a lamp, one may benefit from investigating the usage of the following optical elements to be utilized near the curved surface, in conjunction with light pipes:

1. Aspheric surface

2. Toroidal surface

3. Grid sag lens

4. Extended polynomial surface

5. Extended polynomial lens

These discrete optical elements can be modeled using **Zemax** and their mathematical descriptions can be found in the **Zemax** manual [109].

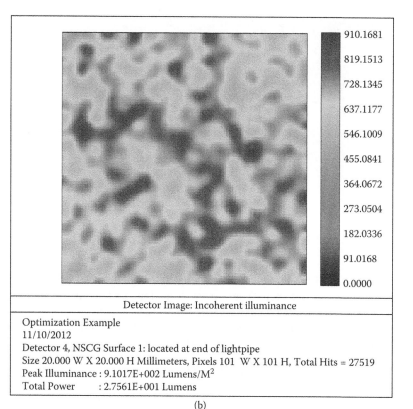

	910.1681
	819.1513
	728.1345
	637.1177
	546.1009
	455.0841
	364.0672
	273.0504
	182.0336
	91.0168
	0.0000

Detector Image: Incoherent illuminance

Optimization Example
11/10/2012
Detector 4, NSCG Surface 1: located at end of lightpipe
Size 20.000 W X 20.000 H Millimeters, Pixels 101 W X 101 H, Total Hits = 27519
Peak Illuminance : 9.1017E+002 Lumens/M^2
Total Power   : 2.7561E+001 Lumens

(b)

Figure 6.20. (continued) (b) Calculated incoherent illuminance distribution in position space observed at the detector that is 43 mm away from the four LEDs shown in Figure 6.20(a). The output shows that the light at the detector is nonuniform, scattered, and that only 6.9% of the luminous flux emitted by the four-LED ensemble is detected.

6.4.3.2 Uniform and Omnidirectional LED Lamp Output Using Tapered Waveguides

In the previous section, we utilized light pipes and compound optical elements to provide the beam expansion from LED emitters. Here we shall consider tapered waveguides for that purpose. As in the case with light pipes, seamless termination at the lamp's outer curved surface may be achieved by means of some curved diffractive optical elements in front of the tapered waveguides. In theory, it may be possible to integrate these discrete elements with the waveguides using molding techniques.

Before starting the analysis, let us first discuss what primarily differentiates *optical waveguides* from light pipes. Light pipes are frequently used to direct polychromatic light such as white light, whereas optical waveguides are structures that guide light waves of single wavelength. The guiding behavior in a waveguide is determined by the optical modes it supports and depends on the optical and geometric properties of the light wave media. These properties determine

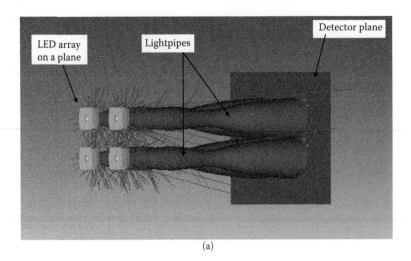

(a)

Figure 6.21. (a) The 3-D schematic showing light rays detected by a 20 mm × 20 mm rectangular detector 43 mm away from four LED sources, which are attached to four light pipes. The traced rays show that the light pipes are not optimized for coupling light from LEDs and thus a good deal of optical power is lost due to radiation at the LED–light pipe interfaces.

how many discrete modes a waveguide can support, which can range from 0 to n, where n is a positive integer. The field of optical fiber communication is largely based on waveguide theories and interested readers are encouraged to learn more from other literature [121,122].

The special interests for using waveguides instead of light pipes to design an LED lamp for general lighting applications are multifold. Waveguide utilization is naturally well suited to LED and laser technologies in terms of achieving enhanced performance, yield, and manufacturability to a great extent, because LEDs inherently are

1. Nearly monochromatic

2. Higher performance devices when their active region or aperture dimensions are kept small

3. Suited to use remote phosphors to produce better lamps [123]

4. Suited to use effective thermal management designs similar to those traditionally used in the electronic and optoelectronic industries

Another attraction for using waveguides in LED lamp designs is that these waveguides will invariably be multimode because their cross-sectional dimensions will need to match the typical practical aperture sizes of LED chips. Multimode waveguides are easier to manufacture than the single-mode counterparts because the former can have larger cross-sectional dimensions and thereby avoid micron

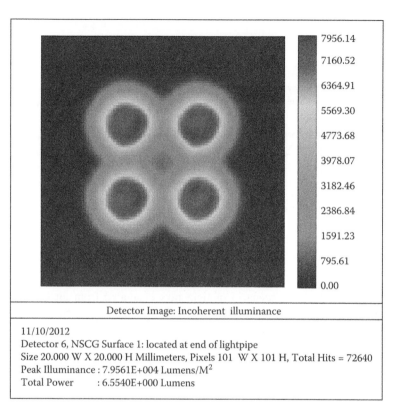

	7956.14
	7160.52
	6364.91
	5569.30
	4773.68
	3978.07
	3182.46
	2386.84
	1591.23
	795.61
	0.00

Detector Image: Incoherent illuminance

11/10/2012
Detector 6, NSCG Surface 1: located at end of lightpipe
Size 20.000 W X 20.000 H Millimeters, Pixels 101 W X 101 H, Total Hits = 72640
Peak Illuminance : 7.9561E+004 Lumens/M^2
Total Power   : 6.5540E+000 Lumens

(b)

Figure 6.21. (continued) (b) Calculated incoherent illuminance distribution in position space observed at the detector that receives light from the four LEDs followed by four light pipes as shown in Figure 6.21(a). The illuminance output shows that the light from the light pipes stays well confined with the detector capturing 16% of the luminous flux emitted by the four-LED ensemble.

or submicron level precisions. They are also easier to align with their corresponding LED chips because larger misalignment tolerances can be accepted. In contrast, alignment of single-mode lasers with single-mode fibers, which are actually single-mode waveguides, is notoriously difficult.

The disadvantage of using multimode waveguides is that many guided modes will be supported by the structure, which can lead to multimode interference and cause nonideal light distribution within the waveguides. The methods to minimize such occurrence is to keep the dimensions of the waveguides and hence the LEDs somewhat *small* and to create the condition so that the fundamental mode can be preferentially excited. The tapering and bending of the waveguides must be fairly adiabatic to keep the fundamental mode dominant throughout the guiding structure [124–126].

Although a rather large size of 1 mm × 1 mm has become a standard for high-power white LEDs, the primary reason for such industry convergence had to

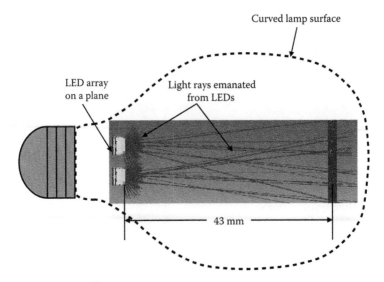

Figure 6.22. A realistic scenario of light rays emanated from an LED array inside a lamp housing is generated using Zemax; the lamp's external configuration resembles that of an Edison-type lamp with a curved outer surface. This illustration shows that not many of the light rays from the LEDs will hit the curved surface orthonormally. The radiation from such a lamp design will not be as multidirectional and uniform as that from an actual Edison-type incandescent lamp.

do with manufacturers trying to stretch the exit aperture window size in order to maximize the total luminous flux from a single SMT LED while maintaining its long lifetime. Utilization of fewer LED emitters in a lamp or luminaire has also been perceived to be economical. However, lasers and LEDs are more efficient and durable if their dimensions are kept small and are operated with much lower electrical power than the typical 1 W per LED chip. This would also be favorable to increasing the wafer-level yield and would lead to more single devices per wafer.

Let us now consider an analysis where adiabatic tapered waveguides are used in a configuration similar to that depicted in Figure 6.24. Here we use six LEDs, emitting green light at 0.55 μm, on an *XY* plane, followed by six waveguides. For simplicity, we only consider that the waveguides all terminate on a plane and not on a curve. We simulate four cases with variable waveguide taper dimensions to show that light distribution at the output plane can be made uniform with tapered waveguides seamlessly joining at the output plane. The simulations are performed using the commercial OptiBPM software from *Optiwave Systems Inc.* [127].

In Figure 6.25, we present the radiant field distribution at the output plane for waveguides without any tapers. In this case, the output plane looks much like the input plane since the ideal and lossless waveguides only propagate the input light from LEDs without broadening them.

In Figure 6.26, we present the radiant field distribution of the case using waveguide tapers in one transverse direction (i.e., along the *X*-axis) only. Figure 6.27 shows the case for waveguide tapers in both transverse directions (i.e., along

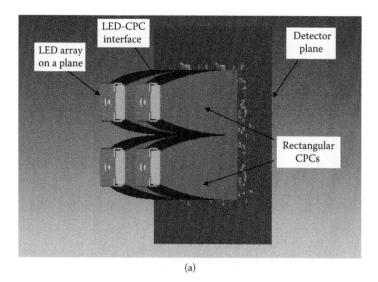

(a)

Figure 6.23. (a) The 3-D schematic showing light rays detected by a 20 mm × 20 mm rectangular detector 10 mm away from four LED sources attached to four rectangular compound parabolic concentrators (CPCs). The CPCs are optimized for coupling light from LEDs and thus almost no optical power is lost due to radiation at the LED–CPC interfaces, which otherwise would have appeared as scattered light at this interface.

X- and *Y*-axes)—however, with finite gaps between the waveguides at the output plane. Finally Figure 6.28 shows the case for tapers in both directions seamlessly joining the waveguides at the output plane.

Figures 6.25 through 6.28 demonstrate how light distribution at the output surface can be made uniform and smoothly varying with seamless joining of tapered waveguides. Figure 6.28 shows that a little more light is concentrated in the middle waveguide compared to the others. A similar effect was also seen in the thermal simulation in Figure 3.10 in Chapter 3. In a discrete array, the center element receives a little more power from the neighboring elements due to symmetry and thus shows a slightly higher concentration of power compared to the others. (The reader was encouraged to seek this explanation in Chapter 3.)

For simplicity, waveguide and LED dimensions are kept very small in the analysis. The interested designer is encouraged to get similar results by increasing these dimensions and preferentially excite the fundamental mode for each tapered multimode waveguide with light launched from each LED.

It is important to recognize that light launched into waveguides should strictly be coherent for ideal waveguiding. LED light output is, however, incoherent. But as long as waveguides are fairly short and LED light is nearly monochromatic, the analysis presented here can be viewed as a realistic approximation. Typical LED lamps will need waveguides (or light pipes) that are *only* several centimeters long, which is rather small and hence appropriate for avoiding severe modal

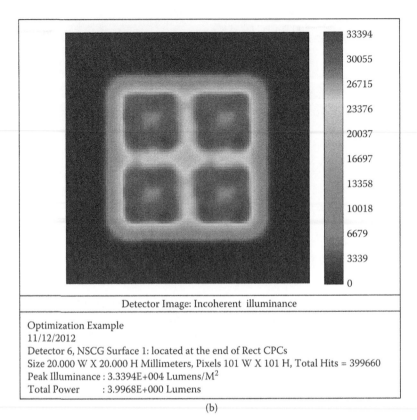

Detector Image: Incoherent illuminance	
Optimization Example	
11/12/2012	
Detector 6, NSCG Surface 1: located at the end of Rect CPCs	
Size 20.000 W X 20.000 H Millimeters, Pixels 101 W X 101 H, Total Hits = 399660	
Peak Illuminance : 3.3394E+004 Lumens/M^2	
Total Power : 3.9968E+000 Lumens	

(b)

Figure 6.23. (continued) (b) Calculated incoherent illuminance distribution in position space observed at the detector that receives light from the four LEDs followed by four rectangular CPCs as shown in Figure 6.21(a). The illuminance output shows the light from the CPCs stays well confined and uniform, with the detector capturing nearly 100% of the luminous flux emitted by the four-LED ensemble.

dispersion effects. As a relevant comparison, LEDs have been successfully used as light sources to send optical signals down multimode fibers over *many* meters, demonstrating negligible modal dispersion effects [128].

The analyses in this chapter have been presented in a simple and systematic manner, which led up to the novel concept of using light pipes and waveguides to construct LED lamps. Such a design concept may be an expensive proposition for general-purpose lamps at this time due to complex manufacturing that may be required. However, without such uniform broadening of LED light output that can be achieved with light pipes and waveguides as shown in this section, the illumination quality will remain poor for a majority of today's lighting applications. The waveguide and light pipe concepts, as well as the analysis presented here, are intended also to provide some important understanding of what type of light manipulation would be necessary from LED chips in order to create broadened

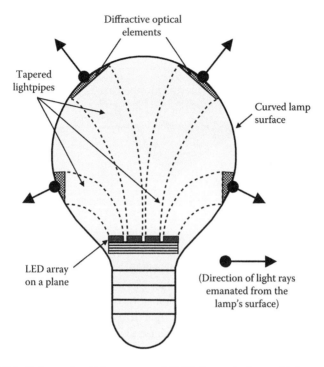

Figure 6.24. Schematic of the proposed LED lamp design, which uses tapered light pipes (can also be waveguides) of different lengths and bends to ensure that the light pipes' optical axes terminate at the lamps outer curved surface orthonormally. Suitable diffractive optical elements may be needed to ensure gradual optical transition at the lamp's exit surface. The illustration uses the external contour of an Edison-type lamp. The radiation from such an LED lamp design is expected to be very multidirectional and uniform. The schematic is not drawn to scale.

light with equivalent uniformity generated by incandescent and fluorescent lamps. This may also help engineers avoid designs that will lead to lamps illuminating inadequately and nonuniformly for many important lighting applications.

Lastly, it should be reemphasized that the waveguide or light pipe design concept *is* beneficial in many important respects. These include the ability to use smaller LED chips, lower input electric power, remote phosphors at the lamp's outer surface, and—last but not least—the ability to utilize many traditional thermal management schemes and technologies that the electronic and optoelectronic industries have long established. These thermal management methods are compatible with those analyzed and simulated in Chapter 3. Furthermore, because light pipe and multimode waveguide technologies are already utilized commercially to some extent [129,130], it is likely that, in the near future, both light pipe and multimode waveguide technologies can be further developed in polymer and poly(methyl methacrylate) (PMMA) materials, which may lead to substantial cost reduction through technology enhancements and high-volume manufacturing.

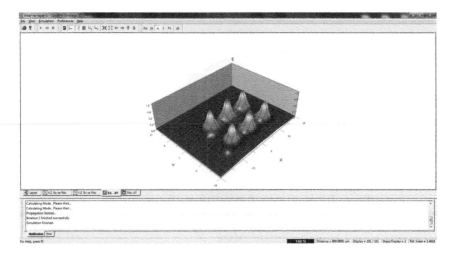

Figure 6.25. Calculated radiant field distribution in position space observed at the output plane that receives light from six LEDs followed by six straight waveguides without taper. The field distribution at the output plane is the same as that of the input plane, which represents the light distribution generated immediately in front of the six LEDs. This is true because the waveguides are ideal and lossless in the simulation.

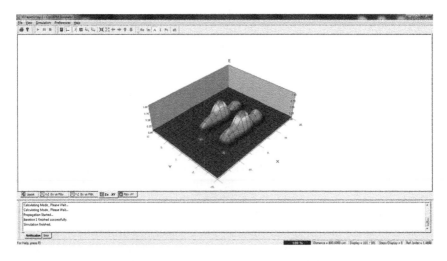

Figure 6.26. Calculated radiant field distribution in position space observed at the output plane that receives light from six LEDs followed by six waveguides with tapers broadening only in the X-direction.

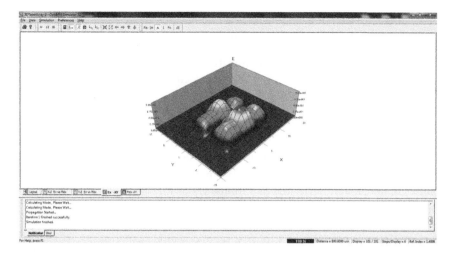

Figure 6.27. Calculated radiant field distribution in position space observed at the output plane that receives light from six LEDs followed by six waveguides with tapers broadening in both X- and Y-directions, but with finite gaps remaining between adjacent waveguide apertures at the output plane.

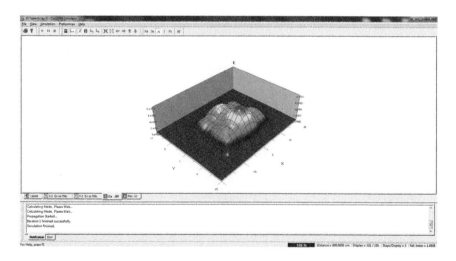

Figure 6.28. Calculated radiant field distribution in position space observed at the output plane that receives light from six LEDs followed by six waveguides with tapers broadening in both X- and Y-directions, with no gaps remaining between adjacent waveguide apertures at the output plane. The tapered waveguides seamlessly terminate at the output plane, producing uniform light distribution.

6.5   Experimental Verification of Expected LED Light Distribution Properties

Thus far, we have discussed LED illumination properties qualitatively and investigated their light distribution behavior using electromagnetic theory and optical simulations. Here we verify some of those properties experimentally.

6.5.1   Measurement and Comparison of Downlight Lamp Data

Several sets of goniophotometric measurements were performed to compare the light distribution from incandescent, compact fluorescent, and LED lamps. In the first set of experiments, the illumination profiles generated by these three categories of lamps were measured in a formal residential dining room at the table top. These were dimmable downlight-type lamps that are meant to illuminate the area preferentially directly below the lamp. Figure 6.29 shows the photograph of the three lamp samples used for the first set of downlight experiments, which are INC-S4, CFL-S3, and LED-S2. (Note: Other photometric properties of CFL-S3 and LED-S2 were presented in Chapter 4.)

The experimental setup for evaluating the three lamps was similar to that described in Section 4.3.1.2 for task lamp measurements. The lamps were placed in a hanging luminaire at the center of the dining table at a height of 30.5 in. above the table top. The table surface represented the XY-plane, which was subdivided into a (17×6) rectangular grid, with grid spacings of $\Delta x = 2.75$ in. and $\Delta y = 6.00$ in. The origin (0.00, 0.00) of the plane was directly below the lamp at the center of the table. The illuminance values were measured at each grid node using a Konica Minolta CL-500A meter. The illuminance maps for the three lamps are plotted in Figures 6.30(a), 6.30(b), and 6.30(c).

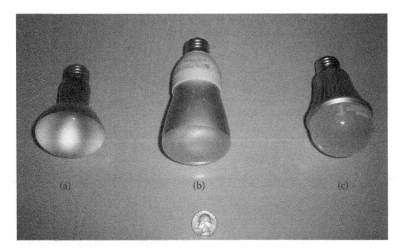

Figure 6.29. Photograph of (a) a 50 W reflector-type incandescent lamp (sample INC-S4); (b) a 45 W equivalent reflector-type compact fluorescent lamp (sample CFL-S3); (c) a 45 W equivalent retail LED downlight (sample LED-S2). The US quarter coin in the photo provides the size reference for the lamps.

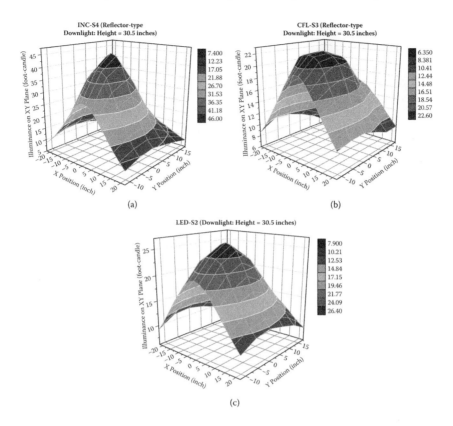

Figure 6.30. Illuminance data taken on a dining table with a lamp height of 30.5 in. for (a) a 50 W reflector-type incandescent lamp (INC-S4); (b) a 45 W equivalent reflector-type CFL (CFL-S3); (c) a 45 W equivalent LED downlight (LED-S2). INC-S4 provides more than twice the foot-candles over the same surface area with greater uniformity compared to the CFL and LED counterparts that are both rated for a slightly less *equivalent* power.

Although the *equivalent* electric input wattage is higher for the incandescent lamp, it can be scaled down properly to make valid comparisons among the three lamps. Hence, utilizing a 10% downscaling of INC-S4, the data in Figure 6.30 confirm that the illumination from the LED replacement lamp is much dimmer and does not spread adequately or uniformly over the 4 ft × 3 ft dining table compared to its incandescent counterpart. In the previous section, these properties were predicted for LED lamps that use discrete arrays from a planar surface. Interestingly, the CFL lamp data in Figure 6.30(b) show a nonmonotonic light distribution as one would expect from a lamp with a spiral surface.

6.5.2   Measurement and Comparison of Ambient Lamp Data

In the second set of experiments, the illumination profile generated by an omnidirectional incandescent lamp and its claimed LED replacement were measured on an informal residential kitchen table top. These were ambient lamps that are

Figure 6.31. Photograph of (a) a 60 W equivalent retail LED lamp from Osram Sylvania (sample LED-S5), which has several discrete LEDs placed on tilted surfaces around the center line of the lamp; (b) a 60 W incandescent lamp (sample INC-S1). The US quarter coin in the photo provides the size reference for the A19 lamps.

meant to provide broader illumination than those in the previous experiment. Figure 6.31 shows the photograph of the two samples, LED-S5 and INC-S1, which are both dimmable. (Note: Other photometric properties of INC-S1 were presented in Chapter 4.)

The experimental setup for evaluating the 60 W incandescent and its LED replacement was similar to that just described. The lamps were placed in a hanging luminaire at the center of the kitchen table at a height of 30.25 in. above the table top. Figure 6.32 shows the photograph of LED-S5 being lit while in the

Figure 6.32. Photograph of LED-S5 as it illuminates from a hanging luminaire at the center of a kitchen table.

fixture, depicting the clusters of LED emitters placed at the bottom of the lamp as well as on tilted surfaces on the sides of the lamp.

In this experimental setup, the grid dimensions, x- and y-spacing, and the origin reference were the same as those in the previous example. The measured illuminance maps for the INC-S1 and LED-S5 are plotted in Figures 6.33(a) and 6.33(b) respectively.

Note: The illuminance plots in Figures 6.30 and 6.33 show some asymmetry along the Y-axis because more rows of data were collected in the +Y direction than in the −Y direction. This occurred because of the constraint imposed by the long length of the Konica Minolta CL-500A meter, which is 6 in.

The experimental results shown in Figure 6.33 confirm that an LED lamp designed with some discrete LEDs placed on tilted surfaces to replace a 60 W ambient incandescent lamp is still unable to match the incumbent's uniformly broad spatial flux distribution that gradually tapers off from the central region. Although the LED replacement lamp generates a much larger peak value at the center, the peak flux diminishes very sharply around the center as opposed to that observed for the incandescent lamp. Such pronounced peak flux levels over a small region are prone to glare and nonuniform object illumination. Finally, it is important to mention that after 2 hours of measurements, the surfaces all around *both* lamps were equally and extremely hot, despite the claim that the frontal surfaces of LED lamps tend to be cool. This may be a result of a compromised thermal management design needed to be incorporated in the LED lamp with emitters placed on tilted surfaces along the sides.

6.5.3   Measurement and Comparison of 3-D Goniophotometric Lamp Data

Let us now investigate certain comprehensive three-dimensional (3-D) gonio-photometric data for incandescent, CFL, and LED lamps measured with a RiGO-801 system from Techno Team. This system was described in Chapter 4, Section 4.2.4.5. The LID data for the three lamp categories were measured by gradually scanning a high-resolution 3-D imaging camera all around the samples. Figures 6.34, 6.35, and 6.36 show the LID data for LED, incandescent, and CFL, respectively, plotted in terms of candelas (cd) in the same-scale XYZ coordinates for comparison. The XZ-plane shows the superimposed angular grid for that plane.

The rated power for the incandescent lamp in Figure 6.35 is 90 W, whereas the rated power for the LED lamp in Figure 6.34 is 12 W. Although the three lamps in these figures have *nonequivalent* electrical input power, we can scale their input powers with respect to their individual luminous efficacy; based on today's average conversion scenario, one may equate a 12 W LED to a 60 W incandescent. By this measure, one can scale down the results in Figure 6.35 to 66% to achieve the equivalency of the LED data shown in Figure 6.34. This agrees with scaling

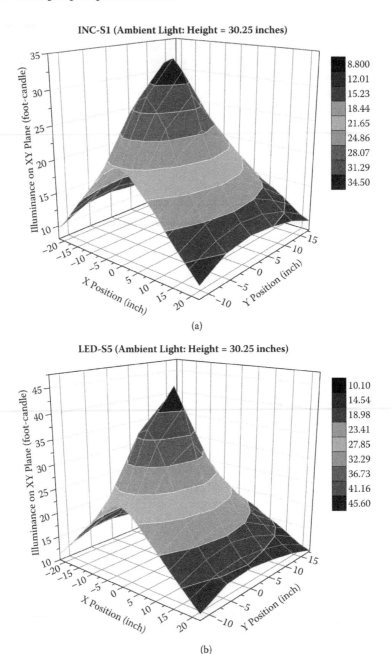

Figure 6.33. Illuminance data taken on a kitchen table with a lamp height of 30.25 in. for (a) a 60 W ambient incandescent lamp (INC-S1); (b) a 60 W equivalent LED (LED-S5). The LED replacement lamp provides higher peak power while the incandescent counterpart provides broader and more uniform or gradually tapering illuminance distribution.

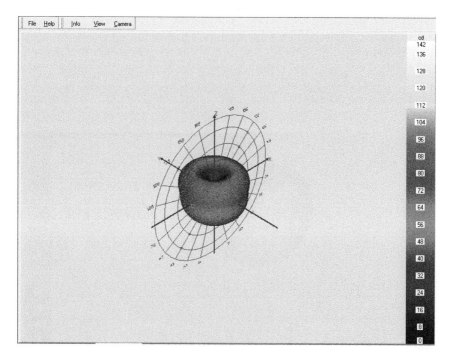

Figure 6.34. (See color insert.) The 3-D plot of the measured LID data for LED-S1 (12 W) using a RiGO-801 system. The graph shows the lamp's luminous intensity distribution in **XYZ** coordinates (in position space), which also shows the angular space grid in the **XZ** plane. The LID data of the LED lamp is largely asymmetric with much of its light distribution remaining in the lower hemisphere where the z-values are all negative. (Data taken by Techno Team staff.)

the incandescent lamp's peak candela power of 108 (see Figure 6.35) downscaled 66% to yield 72 cd, which is close to the same peak power seen in the data for the LED lamp in Figure 6.34.

While their peak powers will be understandably equivalent at the same electrical input power, one can still see that their SFDs converted from the LID data presented here will be quite different. As seen in Figure 6.35, the LID of the incandescent lamp is more gradually distributed in 3-D space over smoothly varying curved contours, covering a much larger volumetric space with broad range of LID values. This type of angular flux distribution is expected to generate a very uniform object illumination with natural shadows, in particular for curved objects. It is also favorable to illuminating a larger volumetric space omnidirectionally since this type of light source will provide a larger integration of flux in space because a continuous set of points from the curved LID profile will contribute to the total luminous flux at locations away from the source.

In contrast, as seen in Figure 6.34, the LID from the LED lamp (LED-S1) does not have a profile covering a broad angular range; and does not show much variation of candle power in 3-D space, where the LID values remain

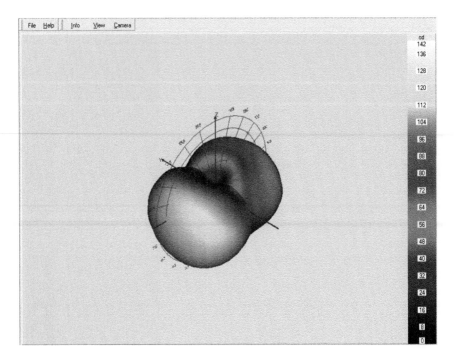

Figure 6.35. (See color insert.) The 3-D plot of the measured LID data for a standard 90 W incandescent lamp using a RiGO-801 system. The graph shows the lamp's position space LID data in **XYZ** coordinates, which also shows the angular space grid in the **XZ** plane. The lamp's LID is spread quite symmetrically over 4π(sr) in both upper and lower hemispheres, showing gradual LID variation over a broad range of values. (Data taken by Techno Team staff.)

It is perhaps a good time now to make an attempt to highlight the illumination behavior of lamps in general with the goal to understand LED illumination further. For that purpose, let us utilize some findings from the field of imaging science. The efforts in determining the shape of an object from image intensities by B. K. P. Horn revealed that, "for a given illumination, the brightness or intensity distribution in the image of an object is related to its surface shape" [131]. He provided a direct relation between the intensity distribution captured in an object's image to its own surface shape. Here we contribute a corollary to this theory:

"For any finite and uniform luminance level, the **luminous intensity distribution** generated by a light source is related to its surface shape." (By finite and uniform luminance level, it is meant that the light source is composed of an optically homogeneous medium of a certain regular size.)

Remark: **Luminous intensity distribution** *of a light source is what generates its "illumination"; that is, a lamp's LID property primarily determines its nature of illumination.*

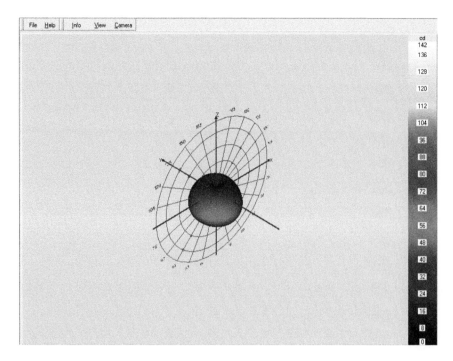

Figure 6.36. (See color insert.) The 3-D plot of the measured LID data for an 11 W compact fluorescent lamp using a RiGO-801 system. The graph shows the lamp's position space LID in **XYZ** coordinates, which also shows the angular space grid in the **XZ** plane. This lamp's LID values are spread somewhat more uniformly over 4π(sr) compared to that of LED-S1 in Figure 6.34. However, this CFL lamp shows higher LID strengths in the lower hemisphere. Its LID variation covers a more limited range compared to that of the incandescent lamp as observed in Figure 6.35 because the incandescent lamp has a much higher equivalent wattage. (Data taken by Techno Team staff.)

nearly the same over a small volumetric region. The light distribution from such a lamp is expected to be useful only for illuminating flat objects close to the light source.

The 3-D goniophotometric data depicted in Figures 6.34, 6.35, and 6.36 essentially support the preceding corollary by showing that the measured LID is different for the light sources of varying shapes. The measured data in Sections 6.5.1 and 6.5.2 also support the postulate that the emitter source shape is responsible for generating a certain illuminance level and its distribution. The simulation, theory, and measurements presented in this chapter all consistently establish that, while current LED lamps are suitable for task lighting and proximity illumination, they are not yet desirable for ambient lighting in the way that the Edison lamp or a CFL is suitable for that use.

7 LED Replacements for Incandescent and Linear-Tubular Fluorescent Lamps

7.1 Introduction

The two most widely used and recognized lamps are still the household incandescent and the linear-tubular fluorescent lamps, which are often called the Edison lamp or bulb and the linear fluorescent lamp (LFL), respectively, in the United States. As discussed in Chapter 5, many countries are trying to make the Edison bulb obsolete for most applications in order to reduce energy consumption. While fluorescent technologies currently provide substantial energy savings over the incandescent counterparts for residential and commercial lighting uses, LED replacement lamps are increasingly viewed as practical contenders for the next generation.

However, the studies in the previous chapters reveal that there are technological as well as some inherent limitations of LED lamps. These must be resolved to a great extent before they can succeed in becoming the dominant lamp choice for consumers. Here, in Section 7.2, we first present certain cost and energy consumption comparisons of various household lamps in order to assist the evaluation of the future home lighting adoption landscape. In the subsequent sections of this chapter, we focus on the LFL since this lamp category currently generates a substantial portion of the consumer lighting industry's revenue; in Section 7.4, we present an improved design for an LED tubular replacement lamp using the novel concept introduced in Chapter 6. Finally, in Section 7.5, we compare some measured data from several LED tubular and LFL luminaires and draw some conclusions.

7.2   Household Lamp Comparisons and Replacement Considerations

Even as phase-out policies are being implemented in many countries, incandescent lamps still remain very popular among many people for household lighting applications. Consumers are swayed by their illumination quality and low retail cost. In addition, many working people only use electric lighting at home during the dark hours and hence view the energy saving argument as only secondary. As the phase-out becomes more and more restrictive, consumers will be driven to use more energy-efficient lamps. Nevertheless, both consumers and lighting industry professionals will surely benefit from recognizing certain facts about cost, quality, personal safety, and other environmental aspects of different lighting solutions.

7.2.1   Why Does the Edison Incandescent Lamp Remain Popular?

Despite their low energy efficiency, in general incandescent bulbs provide very desirable ambient illumination in terms of color, light distribution, and brightness. A regular incandescent lamp, whose measured LID data resemble the kind shown in Figure 6.35, basically sets the illumination standard for other ambient lamps. Such a lamp has a radiative feel to it like the sun; its color, radiation balance, and evanescent trickling are all splendid. Uniform radiation is produced all around the bulb and appropriate luminaires can direct the light where desired.

Incandescent lamps do not produce glare and are continuously dimmable over their entire luminous range using TRIAC (triode for alternating current) switches off the main supply voltage, which remains as the infrastructure in many homes in the United States. Unlike LFLs, they can be easily installed in an "Edison socket" screw base and can be placed at any height and virtually anywhere in a room with a reasonably small portable luminaire without requiring built-in electrical sockets or fixed luminaire housing in proximity. These features make them very suitable for residential and certain commercial applications.

Although CFLs are made to fit into the same incandescent lamp or Edison socket and can provide equivalent luminous flux in broad angular directions over wide spaces, their color and light distribution properties fail to meet the stringent quality requirements of certain users who prefer to pay for higher energy costs for incandescent lamps. Despite the recent improvements in CFL technology, many users are still dissatisfied with their color, color stability, and dimmability features. LED replacement lamps are also facing similar challenges to meet the features of the standard incandescent lamps, with an additional barrier: generating light distribution over broad angular directions that can spread over larger volumetric space. However, the LED lighting industry has demonstrated unprecedented improvements in the past few years and it has potential to improve much further.

7.2.2   LED Replacement Opportunities for the Edison Lamp

Although artists, photographers, and filmmakers understand and appreciate high illumination quality, such affinity is not typical among the general population. Further, there is usually not a pressing need for implementing high-quality illumination for quick tasks and other relatively short-duration experiences. Reduction of energy consumption is far more essential for a large number of applications where quality could be compromised. Thus, both CFLs and LEDs currently stand out as creditable lighting solutions for many ambient and household applications. It is therefore meaningful to compare CFLs and LEDs as replacement options for the Edison lamp.

The CFL and solid-state lighting (SSL) industries have both recognized that, to succeed, a retrofit lamp must be available for America's most popular 60 W incandescent light bulb, which comes with an Edison socket screw base. In late 2010 and early 2011, 60 W, A-line LED replacement lamps from Philips and Osram Sylvania became available in major retail stores and, according to recent announcements, their corresponding 100 W incandescent equivalent lamps were to be available in December 2012 and in early 2013 respectively [132]. Various announcements as of April 2013 confirm their retail availability. While such product availabilities are notable milestones, the questions remain whether consumers are buying these lamps in appreciable quantity and if they are happy with the performance given that price tags are still between $30 and $50 per lamp depending on the wattage consumption.

The answers will unfold over time because a good number of users must invest in the lamps and, by doing so, will experientially conclude whether the LED lamp retrofits are satisfactorily comparable to the 60 W incandescent lamps and superior to the CFL alternative. The sales results will also establish if the Edison-base LED lamps' claimed advantages as higher efficiency, longer life span, and eco-friendliness validate the higher up-front cost. Consumer choices and satisfaction will also depend on effective counteraction against CFLs' mercury content and recycling costs that elevate the full-life cost of a CFL. In the state of Minnesota, for example, CFL recycling costs can average as much as $2 per unit. In addition, consumers are not yet convinced of the benefits of CFLs because of their unsuitability for outdoors, slow starts, flickering, poor color rendering index (CRI) and correlated color temperature (CCT), and they usually fail to last as long as the package claims. Though such grievances create an opportunity for LED lamps, the competition among the three technologies will likely continue for some time, especially in the United States.

7.2.3   Energy Savings Comparisons of LED, CFL, and Incandescent Lamps

While it may take some time before residential consumers can reach their own conclusions, here, we provide a comparative analysis of the common Edison-base CFL, current retail LED retrofits, and incandescent lamps with respect to cost benefits.

Philips and Sylvania LED retail lamps are among the primary options for the 60 W incandescent lamp, both of which have fairly similar performance ratings. They produce dimmable light, with a maximum output of approximately 800 lm at nominal electrical input wattage of approximately 12 W. The light output is more "ambient" compared to most other LED lamps because both produce light with discrete LEDs from multiple tilted surfaces rather than one flat plane. Although prices have been somewhat reduced in more recent times, their original retail price was nearly $40. In contrast, the 60 W incandescent-equivalent, non-dimmable CFL, also consuming approximately 12 W, costs less than $5.

Although the Philips and Sylvania LED replacement lamps have similar total lumen output to that produced by the ubiquitous 60 W incandescent bulbs, the light distribution they produce is not as uniform and omnidirectional as that from incandescent lamps. This was observed for the Sylvania lamp (see Figure 6.31a) as shown in the comparison data in Figures 6.33(a) and 6.33(b). The Philips ambient LED has an unconventional shape, which helps it surpass most traditional LED bulbs in light dispersion characteristics. Its crowned dome appears to cover discrete LEDs mounted on vertically tilted surfaces, allowing light to emit mostly from the sides rather than from the top and bottom (see Figure 2.8). Therefore, its light distribution is also not expected to be as omnidirectional as that of incandescent and CFL counterparts. Both types of LED lamps appear to have unconventional heat-sink fins that handicap ambient illumination. The Philips lamp has an added unconventionality: its French-vanilla colored globed cover—a necessary characteristic that led Philips to imprint "white light when lit" on the lamp itself. The yellowish-cast globe results from remote phosphor coating on the bulb's interior surface, which converts the blue LED chip to a warm white light.

With respect to color temperature, both retrofit lamps' CCT is near 2700 K, which appears as relatively warm, yellowish/white light similar to that produced by the incandescent counterparts. The Philips retrofit LED lamp has a CRI of 80, which is comparable to a CFL lamp, while the Sylvania version's CRI is 90. Both types of LED retrofit lamps have a rated life span of 25,000 hours, which far exceeds that of the CFL replacement counterparts. But the superior advantage of the LED-based lamps is that they do not break easily and are free of both mercury and lead and thus are more eco-friendly than CFLs.

The current cost of LED replacements for America's most popular 60 W incandescent bulb is steep. But do the advantages actually offer savings in the long run? Although the jury is still out, the comparative analysis presented in Table 7.1, using typical numbers from 2010, should provide some food for thought.

Table 7.1 was formulated at an average energy cost of 11.45 cents per kilowatt-hour [133]. It shows that LED lamps save money over incandescent but not CFL lamps. One can brighten the LED statistics by projecting that LEDs will perform well throughout the forecasted 22-year life span, which nevertheless

Table 7.1. Long-Term Cost Comparison Data of Common Household Ambient Lamps

Lamp Type	Life Span (Hours)	Wattage (Watts)	Up-Front Lamp Cost ($)	Average Lamp Cost Per Year ($) (Always On)	Energy Cost Per Year ($)	Total Cost Per Year ($)	Average Lamp Cost Per Year ($) (on 4 h/day)
LED	25,000	12	40	14.03	12.04	26.07	4.35
CFL	15,000	12	4	2.34	12.04	14.38	2.40
Incandescent	800	60	0.5	5.49	60.18	65.67	10.95

requires experiential backing. On the reverse side, if one burns out an LED lamp, moves it, or misplaces it, then the short-term loss is roughly $40. Finally, the figures will obviously vary if any of the lamp types are bought in bulk, securing discounts.

Because fluorescent technologies offer valuable alternatives for household lamps as well as for many commercial lighting applications, it is important to remember that the energy and cost savings brought forth by LED replacement lamps should not be projected based only on the LED versus incandescent scenarios. The CO_2 emission reduction is sometimes calculated based on this unilateral comparison, which usually applies a 75%–80% energy savings for LED replacement lamps over their incandescent counterparts [97,134]. Since fluorescent lamps are already widely used for overall global lighting, the energy savings and therefore greenhouse gas reduction are not expected to be as optimistic as usually projected unless LED technology proves to be substantially more energy efficient than fluorescent technologies. Nevertheless, such LED technology improvements will duly come with trade-offs expected from theoretical limits and perhaps higher up-front costs.

7.3 Why Are Linear Fluorescent Lamps Popular in Commercial Buildings?

Linear tubular fluorescent lamps, or LFLs, are widely used in commercial lighting, which constitutes a large majority of overall lighting usage worldwide. According to the International Energy Agency (IEA), in 2005, the average light consumption and energy consumption per person for *commercial application* were 44% and 40% of total lighting usage globally, which represented the largest sector in the lighting industry in both categories [135]. Industrial lighting came in second in terms of light consumption (28%) but not in energy consumption (20%) because, while residential light consumption constituted 14% of light consumption, it captured a larger percentage (30%) in the energy consumption category [135].

The same study by IEA also identified that, in 2005, LFLs were used to account for 57% of light consumption per person of the total, which also included the Edison lamp (9.5%), tungsten halogen (2%), CFL (4.5%), and HID (high-intensity discharge) (27%) as the other lamp types for residential, outdoor, industrial, and commercial applications encompassing the overall lighting industry. Commercial lighting utilizes 57% of total LFL usage while the remaining is divided between industrial (32%) and residential (11%) lighting. Within commercial and industrial lighting sectors, LFLs dominate with 74% and 62%, respectively, of total light consumption. Unquestionably, all of these data conclude that there is a very wide usage of LFLs, particularly for commercial lighting. IEA has projected that while LFL usage in the lighting industry will drop in 2015 and in 2030, it will still be the dominant luminaire over others. A utilization drop is expected from an increase in the total light produced by future LFLs due to increased luminaire efficiency, room and lighting design, and use of daylight and control systems, as well as introduction of other technologies.

Table 7.2. Energy Consumption by Lighting in Commercial Buildings

Commercial Category	Lighting's Portion of Total Electricity Consumption	Lighting's Electricity Consumption
Retail and service	59%	87 billion kWh
Education	56%	36 billion kWh
Office	44%	86 billion kWh
Health care	44%	27 billion kWh
Food service	30%	15 billion kWh

Source: EIA, 1995 survey.

Note: The electricity intensity (kilowatt-hour per square foot) varies for the categories; the cost of electricity also varies for the categories depending on the total amount of usage.

7.3.1 Features and Benefits of Linear Tubular Fluorescent Lamps

Let us now see why LFLs are so useful. In commercial and industrial environments, *many* varieties of task-oriented lighting are needed over large volumetric space for long periods throughout the day. Prior studies completed by the US Energy Information Administration (EIA) have shown that, in some cases, lighting was the single most electricity-consuming application in many commercial sites that include schools, hospitals, office, and many retail and service buildings in the United States [136]. EIA's 1995 Commercial Buildings Energy Consumption Survey found that lighting constitutes a very significant portion of various commercial applications, as shown in Table 7.2.

Currently, LFLs are the best suited lamps for a majority of the lighting applications in the preceding categories because of their large sizes and high luminous efficacies compared to other fluorescent and gas-discharge lamps. In Chapter 5, Section 5.3.2.3, we briefly discussed that current LED replacement tubular lamps are still unsuitable for illuminating large space from high ceiling heights. As explained in Chapter 6, a larger luminous flux integration can be achieved at various points in space much farther away from a light source if it has a larger surface area that is also curved, compared to those sources that have small, discrete, and planar light emitters such as many LED replacement lamps do. Although the total flux output may be the same for certain types of LFLs and some LED counterparts, the illuminance levels produced by the LED replacements are usually different for illuminating the space of interest. By utilizing lamps that produce the most uniform luminous flux per unit of electric power consumed across the entire desired large volumetric space, one implements the most energy-efficient lighting. Currently, LFLs are superior for this reason for most applications and they also cost much less than the LED replacement lamps.

7.3.1.1 Energy Efficiency of Linear Tubular Fluorescent Lamps

In an effort to reduce the vast amount of electrical energy consumed in the many applications discussed earlier, luminaire engineers have been improving luminous efficacies of LFLs for many years. The improvements unfolded through three LFL generations over which the tube diameter has become smaller. In the 1930s the first generation started with 1.5 in. (38 mm) **T12 lamps;** the subsequent second (1978) and third (1995) generations introduced **T8 and T5 lamps** with 1 in. (26 mm) and 5/8 in. (16 mm) respectively [137]. The classifications are made with "T" representing the tubular lamp shape, followed by the number representing the lamp diameter in eighths of an inch. As such, **T5** lamps have a diameter equal to **five-eighths** of an inch, **T8** lamps have a diameter of **1 (8/8)** in., and **T12** lamps have a diameter of **1.5 (12/8)** in. **T5** fluorescent lamps have the highest luminous efficacy of the three categories, which ranges between 85 and 100 lm/W; efficacy drops for **T8** and **T12** lamps to about 75 lm/W and 60 lm/W respectively [138].

The different LFL types require their own electrical input conditions and therefore are designed with unique lengths and socket pin connections in order to prevent any lamp damages or hazards from circuit mismatches that might occur during installation. Thus, **T5** lamps use a miniature bi-pin base with a smaller electrode spacing compared to those in **T8** and **T12** lamps that use a medium bi-pin base. In order to avoid the usage of **T5** lamps in the older fixtures as replacements, they are made slightly shorter than **T8** and **T12** lamps. For example, a nominal 4 ft **T5**'s length is 45.2 in., whereas it is 47.2 in. for **T8** and **T12** lamps. Retrofitting would require an adapter and changing the ballast inside the lamp fixture [139].

7.3.1.1.1 Ballasts for Linear Tubular Fluorescent Lamps All fluorescent lamps require ballasts to limit the current flowing through the tubes because they represent a load with a negative differential resistance; that is, their resistance drops under a constant-voltage operation, causing the current to increase indefinitely. Without a ballast, the current in the tube would rise to catastrophic levels, spoiling the operation soon after. Fluorescent lamps mainly use two ballast types: magnetic and electronic. Many magnetic ballasts often use large-size inductors and, occasionally, capacitors and transformers to provide current limiting and starting functions in higher wattage lamps, which can generate a good amount of heating and acoustic noise arising from line-frequency hum. While electronic ballasts are more expensive, they use semiconductor microelectronic integrated circuits to provide all necessary functions without generating much heat or noise. They operate at higher frequency and are more energy efficient, which saves money over the long term [140]. Figure 7.1 shows a photograph of an electronic

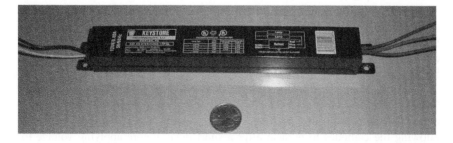

Figure 7.1. A photograph of an electronic ballast for a fixture or troffer that houses two 4 ft **T8** linear fluorescent lamps (LFLs). The US quarter coin provides the size reference.

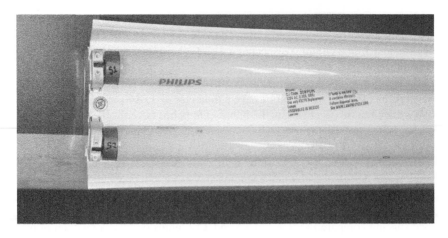

Figure 7.2. A photograph of two 32 W LFLs in a fixture that uses the electronic ballast type shown in Figure 7.1.

ballast for a fixture that houses two 4 ft **T8** LFLs. Figure 7.2 shows the photograph of two 32 W LFLs in a fixture that uses the ballast shown in Figure 7.1.

Advanced electronic ballasts for fluorescent lamps also allow dimming and remote control using networks such as DALI, DMX512, and DSI [141–143]. In order to suit various applications while maximizing energy efficiency at minimal cost, a number of fluorescent lamp ballasts that provide different starting mechanisms are available. These are described next.

7.3.1.1.2 Instant Start As the name suggests, an instant start ballast turns the lamp on almost immediately by applying a very high voltage (~600 V) and avoiding heating the cathodes. It is the most energy efficient among the various ballast types, but limits the number of starts severely. The high-voltage operation sputters oxide materials from the cold cathode surfaces each time the lamp is turned on, thus reducing the lamp's lifetime. This ballast type is most suitable for applications with long duty cycles in which lamps turn on and off infrequently.

7.3.1.1.3 Rapid Start A rapid start ballast achieves a longer lamp life allowing more on/off cycles using slightly more energy compared to the instant start counterpart. It simultaneously heats the cathodes and applies a lower voltage to the lamp, avoiding a cold start. This ballast type is suitable for adding a dimming circuit, which then maintains the heating current while allowing lamp current to be controlled.

7.3.1.1.4 Programmed Start A programmed start ballast operates serially rather than simultaneously by first applying power to the filaments and then allowing the cathodes to preheat after a short delay before applying voltage to the lamps for striking an arc. It is a more advanced version of rapid start and provides the longest life and most starts from lamps. Therefore, it is preferred for applications that demand very frequent power duty cycles, such as in restrooms.

Lamps that utilize these ballasts can provide further energy savings by using motion detector switches or occupancy sensors.

Tubular fluorescent lamp technologies have greatly improved in terms of lamp, luminaire, and ballast designs to effect substantial energy savings. Energy savings result not only from utilization of more energy efficient lighting, but also from reduction of lamp operating times. Thus, incorporation of the most appropriate ballasts for specific applications, usage of automatic controls such as infrared or ultrasonic motion sensors, and daylight harvesting can substantially enhance energy savings for commercial applications. Lastly, because electronic ballasts generate less heat during operation, they also help reduce energy consumed for cooling.

7.4   Linear Tubular LED Replacement Lamp Developments

Many manufacturers in the LED lighting industry see an opportunity in producing LED replacement lamps for LFLs with the outlook of capturing a significant portion of the gigantic market. In the last few years, a large number of vendors have been advertising tubular LED replacement lamps being superior to LFLs in terms of longevity and ease of handling and installation, while offering the same illumination quality with superior energy efficiency. However, in general, the industry is still lacking the practice of quantifying the comprehensive set of lighting metrics discussed in Chapter 4, and thus such claims are often not determined through adequate test results.

7.4.1   Department of Energy CALiPER Tests for Linear Tubular Lamps

In Chapter 4, Section 4.4.3, the importance of careful evaluations by several independent and accredited laboratories was discussed; such labs also need to characterize fully and compare accurately the features of incumbent and LED replacement lamps. In 2010 and 2011, Department of Energy (DOE), collaborating with independent testing laboratories, performed tests on several LED replacement lamps from various manufacturers and compared the results against certain corresponding benchmark LFLs. These tests were performed under the *SSL CALiPER Program,* in *Rounds 11* and *12,* applying an effort to maintain lamp parameter similarities as much as possible [144,145]. The tests have concluded that, while LED technology as well as retail products have improved significantly in the past few years, wide quality variations among different products and manufacturers still exist; further, there were large discrepancies between some manufacturers' claims and actual product performance.

Such a scenario is a result of the fast-evolving LED lighting industry that is not yet mature; many manufacturers tend to bypass thorough design processes for specific applications and reliability requirements and thus the churned products are not altogether suitable for intended current uses. Therefore, the burden falls on buyers and specifiers, who must acquire the knowledge necessary to evaluate and compare products through proper tests. It is also crucial for specifiers to obtain valid LM-79 or equivalent reports from LED lamp and luminaire manufacturers prior to purchase.

In Rounds 11 and 12, several 4 ft LED replacement lamps were tested following LM-79 procedures and their performance were compared with a number of high-performance **T8**-LFLs. These tests showed that, although in some cases LED replacement lamps had higher efficacies ranging from 74–78 lm/W, each lamp nominally produced only half the total lumen output at nominally half the input wattage. This is not surprising because currently a tubular LED lamp surface does not emit in full capacity as an LFL because the former only emits light from, at most, half the surface area due to the appreciable gaps between adjacent discrete SMT modules [146].

Figure 7.3 shows an illustration of a typical retail tubular LED replacement lamp where at least 50% of the lamp's surface is void of light sources. Such a discrete arrangement of LED emitters would still produce only half the lumen output of an LFL even when SMT LEDs are placed all around the tube and not just within a hemispherical surface, provided the LED's and fluorescent lamps' efficacies and power factors remain similar. This would be true for a tubular LED replacement even when light from its top hemisphere is efficiently collected and directed out of the fixture into the ambient space.

The CALiPER tests also showed that the LED lamps with higher efficacies had slightly lower color performances. Most of the tested LED lamps showed significantly less LID strength, uniformity, and angular coverage compared to the LFLs. Such characteristics persisted even in the nonparabolic troffer fixture,

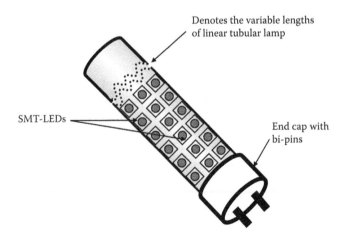

Figure 7.3. An illustration of today's typical commercial LED replacement lamp for linear tubular fluorescent lamps. Current replacement lamps place LED sources directly on the lamp surface with appreciable gaps between the adjacent SMT modules as shown here.

termed as "high-performance" troffer, for which the LFL's light distribution narrowed significantly, deviating from its broad "bat-wing" spread. However, in most practical applications, LID narrowing with a nonparabolic troffer is not useful because a broader and uniform LID spread is more efficient than a more concentrated LID with larger strength. Therefore, in many cases, it is often not practical to seek a nonparabolic troffer design just for the purpose of making the differences between the LED replacement lamps and LFLs smaller in terms of the angular spread in LID, which determines the luminaire spacing criterion.

The CALiPER tests confirmed that the LED replacement lamps are still underperforming LFLs, particularly in the light distribution category for which LFLs are purposefully designed and utilized effectively for large volumetric space lighting. As discussed previously in Chapters 5 and 6, the current LED counterparts fail to compete because they are constructed with small, discrete, directional and flat light emitters and lack appropriate secondary optics that give rise to two deficiencies: (1) low LID strength and (2) LID confined to small regions; *both* of these assertions were confirmed by the CALiPER test results.

Without proper scaling of LID to match the levels produced by LFLs, LED replacements will not become the primary choice. LED arrays inherently produce concentrated light in receptacles, thus not allowing effective overlapping of luminous flux needed to boost illuminance on surfaces and regions far away from the source. The result is inadequate illumination at farther distances over wider regions despite having higher unit luminance and unit luminous efficacy. Decreasing the luminaire spacing criterion is generally not a good solution because LED's inherent problem with glare would only become more pronounced.

7.4.2   A Novel LED Lamp Design for Replacing LFLs

LED replacement lamps would compete successfully with the incumbents if they could be constructed to provide appropriate LID strengths and distributions while providing higher energy efficiency, longevity, and color quality. In Chapter 6, the concept of utilizing tapered waveguides to broaden light from discrete LEDs was introduced. Here, we apply that concept to a linear LED luminaire structure to design an LFL equivalent in terms of broad and uniform LID distribution while providing an uncompromised passive thermal management scheme that can substantially extend the lamp's lifetime.

7.4.2.1   *Features and Benefits of the Novel LED Replacement Lamp*

Currently, many LED-based replacement lamps for tubular fluorescent lamps place discrete surface-mount technology (SMT) LEDs directly on a cylindrical base; this produces wasteful, nonuniform, and directional illumination that is unsuitable for large-space and high-ceiling applications. They are particularly wasteful when discrete LED modules are placed all around the tubular surface because the emitters in the upper hemisphere produce light upward against the ceiling. The novel LED lamp design [119] described here comprises a number of discrete LEDs mounted on a common substrate, where ideally all the light from each LED is immediately guided and broadened through an adequately long and tapered waveguide that seamlessly terminates at the lamp's semicircularly curved cover surface.

Many such LED-waveguide assemblies can fill the entire curved cover and produce diffused light distribution, resulting in uniform illumination over broad angular ranges, as explained in Chapter 6. The proposed lamp has a D-shaped cross-section where the lamp's flat side is used as a heat-sink base to be placed against the ceiling or some blocking surface, which thereby enables an effective passive thermal management scheme that can substantially extend the lamp's lifetime.

The three-dimensional (3-D) schematic illustration of the lamp's full exterior structure is shown in Figure 7.4. Figure 7.5(a) shows the drawing of the lamp cut along its length in order to illustrate the LEDs followed by waveguides inside the lamp's cover; the enlarged version of the lamp's "cut-end" is shown schematically in Figure 7.5(b). Finally, in Figure 7.6, the two-dimensional (2-D) cross-section of the lamp is shown emphasizing each *tapered* waveguide's seamless termination at the lamp's curved cover. It should be noted that a multimode waveguide's core can be tapered without changing the cylindrical outer shape of the waveguide.

The lamp design concept just described allows the usage of today's widely utilized inorganic LEDs that are small, planar, and monochromatic, such as blue LEDs mounted on a flat base board. The novel design exploits these familiar and mature technologies from the optoelectronic industry to create uniform and broad light distribution that is not typical for LED lamps. The tapered waveguides provide uniform and broadened light output, without creating any dark spots on the surface. Further, they allow the usage of remote phosphor on the lamp surface to generate white light with desirable chromatic properties. However, the real flexibility of this design is that any LED emitter, in die or module form, white or single color (e.g., red, blue, green, etc.), or even OLEDs may be used as light sources.

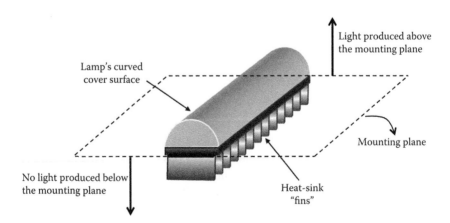

Figure 7.4. The drawing of the novel lamp's external structure showing its placement on a mounting plane. The lamp produces light only on one side of the mounting plane over a hemisphere, while the other side is used to house the protruding or extruded heat sink.

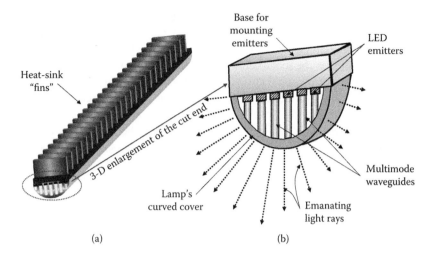

Figure 7.5. The 3-D schematic of the novel lamp showing certain inside details: (a) the illustration of the lamp cut along one end showing the LED emitters inside followed by waveguides; (b) the enlargement of the cut end, which shows that a multimode waveguide follows each LED emitter and terminates at the lamp's cover seamlessly.

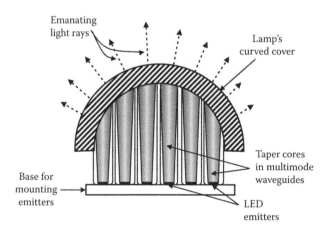

Figure 7.6. The 2-D cross-section schematic of the novel lamp showing the discrete LEDs followed by tapered-core multimode waveguides that seamlessly terminate at the lamp's curved cover. The lamp is expected to produce uniform light distribution in radial directions as shown with dotted arrows at the lamp's curved surface.

Most current linear LED replacement lamps have tubular structures that imitate LFLs with respect to the outer shape. However, this is not necessary because, while LFLs need to utilize the full cylindrical volume structure to generate efficient light using a gas discharge mechanism, LEDs need not use the full cylindrical structure. Generally, LFLs are placed against the ceiling in appropriately shaped louvers in each troffer unit (as shown in Figure 5.16) to direct most of the unwanted upward light downward to illuminate large rooms. Since LED lamps need not utilize a full tubular or cylindrical structure, the light generated by them can be directed only in the space below the ceiling by design. Hence, half a cylinder with a flat top (i.e., a lamp with a "D-shaped" cross-section) would illuminate areas below the ceiling effectively without the need for a recessed design with reflectors. Consequently, the novel LED lamp design incorporates a semicircle or hemispherical cross-section (i.e., half a cylindrical tube that results in a flat surface on one side that does not illuminate). The flat side goes against the ceiling and thus makes it suitable for placing an adequately large heat sink for effective thermal management, which is critical when a large array of high-powered LED emitters is used in a lamp or luminaire.

Based on the theory and simulation presented in Chapter 6, the design concepts described here are expected to provide substantially more uniform and glare-free illumination compared to today's LED-based replacement lamps. As long as the luminous efficacy of LFLs and LED replacement lamps is in the same ballpark, one would still need two of these semicircular LED lamps to replace one LFL in order to produce the same total lumen output. However, unlike most current LED retrofits, which also require two lamps to match the total light output of an LFL, the proposed lamp-pair would rival an LFL in term of generating uniform and broad light distribution. Further, if a single LED emitter's efficacy is doubled in the future, the proposed lamp, as a stand-alone, would be expected to produce the same total lumen output and broad light distribution as that of an equivalent LFL, with only half the input electrical power.

On the other hand, as long as the LED replacement lamps continue the usage of discrete LED emitters leaving dark surfaces on the tubular structure, increased efficacy will only match or somewhat exceed the total lumen output of an equivalent LFL having the same spatial volume; however, they will not match an LFL's desirable light distribution properties. Even if a chip-on-board (COB) or other equivalent technologies are used to reduce the gap between emitters, flat emitters without any appropriate secondary optics will still produce excessive glare and narrow angular light distribution.

7.4.2.2 Overcoming Manufacturing Challenges for the Novel LED Lamp

The proposed novel lamp may face some manufacturing challenges to realize the tapered waveguides or some equivalent light pipes that may also need some end diffractive optical elements as discussed in Chapter 6. But multimode plastic, polymer, or PMMA (poly(methyl methacrylate)) waveguides as well as light pipes, which have been used with LEDs and vertical-cavity surface-emitting lasers (VCSELs), have already been commercial optical components for many years [147]. Therefore, it should be possible to utilize such passive optical platforms to realize any further developments that may be necessary for the proposed LED replacement lamp designs introduced in Chapters 6 and 7.

7.5   Comparison of Measured Data from Various Tubular Lamps

In this final section, we shall investigate some measured performances of several LED and fluorescent linear tubular lamps and analyze their compatibility. Application of relevant photometric measurements and valid comparison criteria are both necessary to assess the overall quality and eligibility of LED replacement lamps for various applications. The DOE CALiPER tests discussed in Section 7.4.1 had justly concluded that buyers and specifiers need to have the knowledge of what lamp parameters and quantitative benchmarks for specific applications are of importance because the vendors' LED lamp ratings are currently not always accurate or comprehensive. The measurement data analyses in this section are intended to provide some guidance for LED engineers, lighting specifiers, and users with respect to important lighting parameters and benchmarks toward considering LED replacement lamps.

7.5.1   Description of Tubular LED Replacement and LFL Luminaire Test Samples

Various photometric measurements were performed on four retail LFLs and 10 commercial LED replacement lamps in an effort to compare their illumination characteristics. All lamps were 4 ft **T8**s, but the LED samples had varying electrical wattage and color ratings, which could not be avoided since the standards are still lacking for SSL replacement lamps. The ratings as listed by lamp manufacturers of the 14 **T8** lamp samples are provided in Table 7.3. The list comprises four retail fluorescent **T8**s (LFL-S1 through LFL-S4) and 10 LED replacement **T8**s (LED-T8-S1, LED-T8-S2, and LEDGREEN-T8-S1 through LEDGREEN-T8-S8). A photograph of two LFL and two LED-**T8** samples together are shown in Figure 7.7. Figures 7.8(a) and 7.8(b) show photographs of two LFLs and 10 LED-**T8** replacement lamps, lit and unlit, respectively.

The LED replacement samples supported AC voltage input ranges covering the different main voltages used in various countries so that the lamps may be adopted globally. The retail fluorescent lamps, LFL-S1 through LFL-S4 were operated in pairs with a retail ballast fixture that had an electronic ballast type such as that shown in Figure 7.1. For the LEDTRONICS samples, the same fixture was used, *but with the ballast removed and the connections rewired according to the vendor's electrical input requirements.* Figure 7.7 shows two LFLs and two LEDTRONICS LED-**T8**s lit using the same retail fixture types from Lithonia. These fixtures may be used as recessed troffers in the ceiling. The LEDGREEN **T8** samples all required their own customized drivers and fixtures, some of which were polarity sensitive.

7.5.1.1   *Retrofitting LED Lamps into Existing Fluorescent Lamp Fixtures*

Linear fluorescent lamp fixtures and troffers have been installed in buildings for many decades, and professional electricians are familiar with their handling and

Table 7.3. Vendor Specifications of Linear Tubular Fluorescent and LED Replacement Lamp Samples

T8 Lamps	Lamp Manufacturer	CCT (Rating)	Total Flux (lm)	Electrical Input (W)	Input Voltage
LFL-S1 and LFL-S2	Philips	5000 K	2850 (per lamp)	32 (per lamp)	120 V-AC
LFL-S3 and LFL-S4	Philips	4100 K	2800 (per lamp)	32 (per lamp)	120 V-AC
LED-T8-S1 and LED-T8-S2	LEDTRONICS	4100 K	1417 (per lamp)	17 (per lamp)	90–290 V-AC
LEDGREEN-T8-S1 and LEDGREEN-T8-S2	LEDGREEN Company	5500 K	2250 (per lamp)	22 (per lamp)	110–277 V-AC
LEDGREEN-T8-S3	LEDGREEN Company	5000 K	1650	15	110–277 V-AC
LEDGREEN-T8-S4	LEDGREEN Company	4000 K	2000	22	110–277 V-AC
LEDGREEN-T8-S5 and LEDGREEN-T8-S6	LEDGREEN Company	5000 K	1750 (per lamp)	15 (per lamp)	110–277 V-AC
LEDGREEN-T8-S7 and LEDGREEN-T8-S8	LEDGREEN Company	5000 K	2100 (per lamp)	22 (per lamp)	110–277 V-AC

Figure 7.7. A photograph of LED-T8-S1/S2 pair (top) and two Philips LFLs (bottom) lit in the same Lithonia fixture types. The ballast inside the fixture was removed for the LED **T8**s.

requirements. However, since LED replacement or retrofit lamps should not use the ballasts (i.e., the entire fixture for LFLs) directly or at all, both users and LED lamp manufacturers respectively face installation as well as engineering challenges. The designers and manufacturers must provide appropriate mounting fixtures and safe electrical power connections for linear LED retrofit lamps so that installers can make the replacement in a straightforward manner.

Since most troffers today include fluorescent ballasts that power the linear lamp mounting brackets, SSL manufacturers use various accommodating approaches, for which standards have not yet been established. Such methods include powering the lamps with the fluorescent ballast, powering the lamps with a built-in driver,

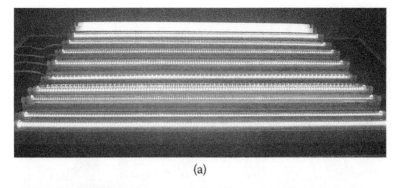

(a)

(b)

Figure 7.8. (a) A photograph of one Philips LFL pair (top) and all 10 LED-**T8** samples lit in corresponding fixtures using the US main supply voltage of 120 V-AC. (b) A photograph of all the lamps in Figure 7.8(a), but void of any applied electrical power and therefore unlit. A Philips LFL-T8 pair is on top followed by all 10 LED-**T8** samples. The bottom-most pair is LEDTRONICS' LED-T8-S1/S2 in a Lithonia fixture with the inside ballast removed.

replacing the ballast with an external driver and rewiring the fixture, or mounting and powering the lamps with detached mounting brackets. As customary in the present industry, all the LED **T8** lamps tested in this study required removing the ballast and rewiring the fixture such that the main input voltage would pass from pins on one end of the lamp to the other. It was discovered that while some retrofit lamp vendors provided schematics regarding rewiring and installation, others had minimal descriptions regarding fixture rewiring requirements. Figure 7.9 shows an external LED T8 driver along with the Lithonia fixture's ballast that was shown in Figure 7.1.

Figure 7.9. A solid-state external driver (top) used in LEDGREEN **T8** replacement fixtures; below it shows the ballast used in Lithonia fixtures for 4 ft LFL-**T8**s. The US quarter coin provides the size reference.

If the ballast is removed and replaced with an external LED driver or rewired for the LED lamp to be directly connected to 120 V-AC line voltage, the fixture is no longer functional for fluorescent lamps. Therefore, it is important to make sure that the users will be satisfied with the LED replacements and will not have to worry about revising their decision. Understandably, the challenges surrounding retrofitting fluorescent fixtures or troffers with LED luminaires raise numerous concerns on cost, safety, methods, labeling, commissioning, and future lamp maintenance including replacements and recycling.

7.5.2   Comparison of Illuminance Data for T8 Lamps

The linear **T8** lamp samples described in Table 7.3 were measured in pairs for linear illuminance profiles directly above the lamp fixtures at two different heights. The purposes of these measurements were to observe (1) the light levels produced by the lamp pairs, and (2) how the near field distributions in position space varied among the lamps. Figures 7.10 and 7.11 show the measured illuminance of the lamp pairs at heights of 2.4 and 3.4 ft respectively.

As one can see in Figure 7.10, the light distributions for the brighter LED **T8** samples (three out of four LEDGREEN pairs) concentrate more substantially directly in front of the lamps compared to the fluorescent counterparts; the brighter the LEDs are, the higher is the concentration toward the middle as seen in the LEDGREEN samples, which show much higher illuminance levels than the other samples at close distances of a few feet. Higher central concentration of luminous flux for arrayed LEDs was also observed in the simulations presented in Chapter 6, where it was demonstrated with two LEDs showing a strong peak or concentrated light in the middle of the detector plane close to the source plane.

When many more discrete emitters are arranged in a 2-D array in the near-field distribution, the light is concentrated more substantially in the middle from the

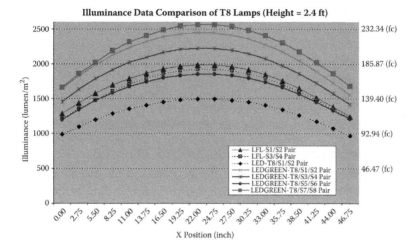

Figure 7.10. The illuminance data measured for 14 tubular lamp samples in pairs using a Konica Minolta CL-500A meter. At a close distance of 2.4 ft directly above the lamps, the LEDGREEN **T8** samples that have more discrete and brighter LEDs show a greater light concentration in the middle than the LFL counterparts do. The "0" X-position on the left side corresponds to one end of the lamp pairs. The left and right vertical axes use lux and foot-candle units respectively.

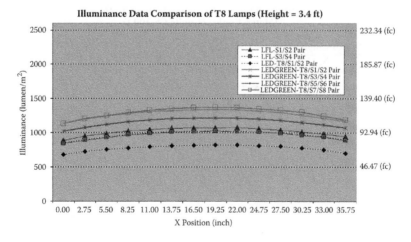

Figure 7.11. The illuminance data measured for 14 tubular lamp samples in pairs by scanning a Konica Minolta CL-500A meter 3.4 ft above the luminaires. The LEDGREEN **T8** samples with brighter LEDs show higher illuminance levels than the other lamp pairs, but their light concentration profile in the middle lessens compared to that seen in Figure 7.10. The "0" X-position on the left side corresponds to approximately 6 in. from one end of the lamp pairs.

contributions of all the LEDs in the emitter ensemble due to symmetry. On the other hand, the near-field LFL light distribution shows less concentrated light immediately in front of the lamp because these lamps, along with the luminaires, are designed to spread luminous flux more evenly over much broader angular ranges. At a greater distance—or height, in this case—the light concentration lessens for the brighter LED replacement lamps in particular, as shown in Figure 7.11. Such lessening of light concentration at the center was also observed in the numerical simulations presented in Figures 6.9–6.11 in Chapter 6.

7.5.3 Comparison of Luminance Data for T8 Lamps

All 14 **T8** lamp samples were individually measured for luminance using a Konica Minolta CS-100A meter. The luminance data for the LFL lamps were straightforward and repeatable and are expected to be accurate since the lamps' emitting surfaces are continuous and they produce uniform angular light distribution. Conversely, obtaining the LED luminance data proved to be more difficult because there were dark spots in between the discrete SMT LEDs that were as large as the emitters themselves. Although attempts were made to minimize errors by focusing only on the emitters, the luminance values measured are still expected to have some inaccuracies associated with them. Nevertheless, the errors should be less than 15% and the data still provide useful qualitative information regarding glare, which was visibly noticeable for the LED replacement lamps. The luminance data for all 14 lamps are plotted in Figure 7.12.

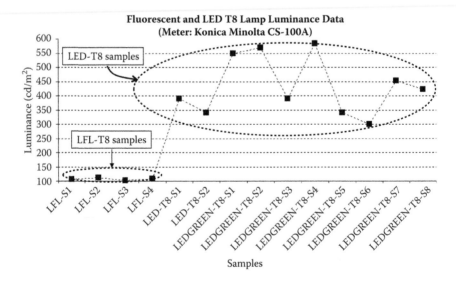

Figure 7.12. The luminance levels measured for four LFL and 10 LED tubular lamp samples using a Konica Minolta CS-100A meter. While the LFL data should be sufficiently accurate, the LED data only represent some average luminance value for each lamp that may be treated as its minimum luminance level.

It is important to note that the real brightness level cannot strictly be characterized for today's LED replacement lamps that have exposed or visible discrete LED arrays. Each measured data point in Figure 7.12 only represents some average value for a particular lamp that may be treated as a minimum floor for its nominal luminance viewed at a distance of a few feet. Based on these arguments, it can be stated that the luminance levels for most of the LED **T8** samples exceed the general comfort level of 200–330 nits when viewed from a distance of several feet. This will be problematic because viewers are expected to be present within such distances from these lamps for many applications. In contrast, when LED lamps are used in billboards or other types of electronic message centers (EMCs), they are typically viewed from distances of between 50 and 500 ft. However, studies have still shown that luminance of LED-EMCs and billboards should not exceed the level of 350 nits [148].

Since currently there appears to be no restriction placed on an upper limit for luminance for lamps of any kind, many LED lamp manufacturers are still attempting to increase their lamp performances by utilizing newer and ever brighter discrete LEDs as the technology improves. While such attempts are not enhancing the light distribution performances, they are producing excessive luminance levels not conducive to the naked human eye. Using a translucent lamp cover will only reduce the luminaire efficiency and lessen the energy efficiency argument for LED replacement lamps and thus is not viewed as an effective solution. A safety or quality standard is likely to be adopted sometime in the future that may limit the luminance levels of general purpose lamps. Similar candidates for this type of concern are the car back or tail lamps that use red LEDs, which are often extremely bright and uncomfortable for the drivers behind.

7.5.4 Comparison of Chromatic Data for T8 Lamps

The color characteristics of the tubular lamp samples were measured using Konica Minolta CL-500A and CS-100A meters. The CIE (x,y) coordinates measured along with luminance for each lamp are shown in Table 7.4. The table also shows the CRI measured for the lamps in pairs.

The spectral waveform data were obtained for the lamps in pairs using a Konica Minolta CL-500A. The data for LFL-S1/S2 and LEDGREEN-T8-S5/S6 pairs are plotted in Figure 7.13. The data for LFL-S3/S4 and LED-T8-S1/S2 pairs are plotted in Figure 7.14. The pair combinations for these comparison plots were chosen according to the paired lamps' similar CCT characteristics and luminous flux levels at the detector.

The spectral distribution seen in Figure 7.13 observed for the LED-T8-S1/S2 pair explains their warm CCT characteristics (rating: 4100 K) because the spectral content is more evenly distributed over the range in the sense that the blue peak is approximately at the same level as the warm peak. Applying a similar justification, the observed spectral behavior for the LEDGREEN-T8-S5/S6 pair in Figure 7.14—showing its blue peak level appreciably exceeding that of the warm peak—results in a higher CCT (rating: 5000 K) than the LED-T8 pair in Figure 7.13. The CRI measured for this LEDGREEN pair was, however, higher than that of the LED-T8-S1/S2 pair as seen in Table 7.4, which resulted from the

Table 7.4. Measured Chromatic Data for Linear Tubular Fluorescent and LED Replacement Lamp Samples

Data Taken: Oct. 15, 2012	CS-100A Data			Data Taken: Oct. 13, 2012 Oct. 14, 2012
		Chromaticity		
Sample	Luminance (cd/m²)	x	y	CL-500A Data CRI (Ra)
LFL-S1	106	0.345	0.361	82
LFL-S2	112	0.345	0.362	
LFL-S3	102	0.381	0.387	82
LFL-S4	109	0.383	0.388	
LED-T8-S1	390	0.386	0.396	69
LED-T8-S2	340	0.381	0.385	
LEDGREEN-T8-S1	550	0.325	0.361	70
LEDGREEN-T8-S2	570	0.326	0.349	
LEDGREEN-T8-S3	390	0.301	0.353	74
LEDGREEN-T8-S4	583	0.377	0.384	
LEDGREEN-T8-S5	340	0.356	0.353	74
LEDGREEN-T8-S6	300	0.336	0.356	
LEDGREEN-T8-S7	453	0.333	0.381	74
LEDGREEN-T8-S8	422	0.332	0.384	

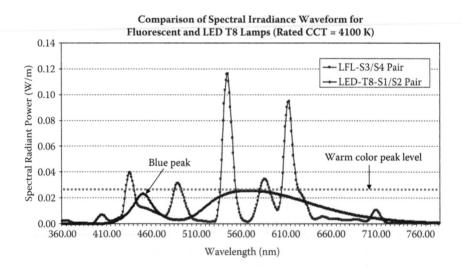

Figure 7.13. The spectral behavior of LFL-S3/S4 and LED-T8-S1/S2 lamp pairs measured using a Konica Minolta CL-500A meter. The blue peak of the LED-**T8** pair sample is less pronounced compared to the warm color peak level, resulting in a warm CCT of 4100 K. The warm peak level is shown using the dotted line as marked in the figure.

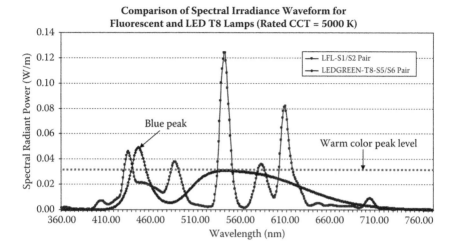

Figure 7.14. The spectral behavior of LFL-S1/S2 and LEDGREEN-T8-S5/S6 lamp pairs measured using a Konica Minolta CL-500A meter. The blue peak of the LED **T8** pair sample is more pronounced compared to the warm color peak level, resulting in a high CCT of 5000 K. The warm peak level is shown using the dotted line as marked in the figure.

higher spectral radiant power levels detected throughout the entire wavelength range for the former pair.

7.5.5   Comparison of Goniophotometric Data for T8 Lamps

We have already learned in prior discussions that current LED replacement lamps do not produce light distributions that are as broad and uniform as those radiated from conventional lamps. This drawback can be most clearly demonstrated by measuring the lamps' LID characteristics comprehensively in three dimensions. In Chapter 6, three-dimensional (3-D) LID profiles in units of candela (cd) were depicted for LED, incandescent, and compact fluorescent LED lamps in Figures 6.34, 6.35, and 6.36 respectively. As discussed in Chapter 4, test procedures required for these types of data are very laborious and expensive, particularly for such large lamps as 4 ft **T8** LFLs and their LED counterparts.

In order to distinguish the light distribution characteristics of certain current LED replacement T8 lamps with those from the incumbent LFLs, we now present the measured 3-D LID data for three sets of **T8** lamps. We shall compare two LED **T8** sets—namely, LED-T8-S1/S2 and LEDGREEN-T8-S1/S2—with LFL-T8-S3/S4 via their LID data presented in Figures 7.15 through 7.17. All three lamp pairs were measured at 23°C after 1 hour of burn-in time to stabilize the lamps.

Figure 7.15 shows the measured LID data of the LED-T8-S1/S2 pair, plotted in terms of candela in *XYZ* spatial coordinates. The sample was oriented at the center of the *XYZ* coordinate system, facing downward in the negative *Z*-direction with its LED emitter plane being parallel to the *XY* plane.

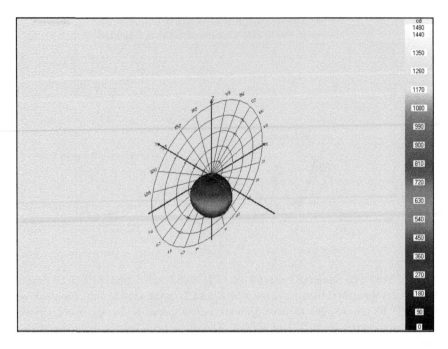

Figure 7.15. The measured LID data for LED-T8-S1/S2 pair using a RiGO-801 system. The 3-D plot shows the lamp's luminous intensity distribution in **XYZ** spatial coordinates, as well as the angular-space grid in the **XZ** plane. The LID profile is a near Lambertian with all of the light distribution residing in the hemisphere in front of the lamp. (Data taken by Techno Team staff.)

Figure 7.15 shows that all the light emitted from the LED T8 lamp pair concentrates in front of the lamp, resembling a slightly modified 3-D Lambertian. The diameters of this near Lambertian vary slightly at various **XY** and **YZ** planes along the **Z**-axis and **X**-axis respectively, compared to those of an ideal Lambertian, because the array of SMT LEDs in the two **T8** lamps contribute a greater amount of integrated flux to the center region than to the outer region due to symmetry as z increases downward in the negative direction, meaning as the distance from the LED emitter plane increases. As expected, these descriptions also hold true qualitatively for the LEDGREEN-T8-S1/S2 pair, whose LID data are shown in Figure 7.16. The DOE CALiPER Round 11 tests discussed in Section 7.4.1 also reported data showing consistent LID patterns of this kind for LED tubular replacement lamps, although those were only in two dimensions representing data corresponding to a single plane. The DOE data were likely for the vertical plane equivalent to the **XZ** plane at $y = 0$, following the **XYZ** coordinate system configuration shown in Figure 7.15 [149].

In contrast to the near Lambertian profile, the measured LID data for the LFL-S3/S4 pair, plotted in Figure 7.17, shows a wider lateral spread along the **Y**-axis on the various **XZ** planes compared to the spread along the **X**-axis on the various **YZ** planes. The reader is encouraged to keep the axis directions and

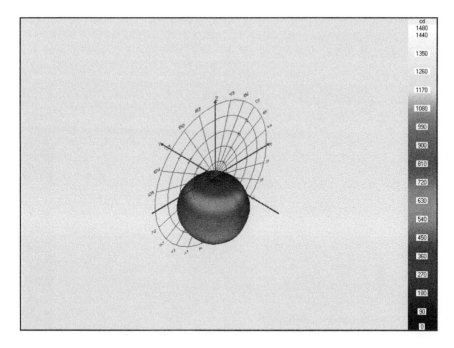

Figure 7.16. (See color insert.) The measured LID data for LEDGREEN-T8-S1/S2 pair using a RiGO-801 system. The graph shows the lamp's luminous intensity distribution in **XYZ** spatial coordinates while showing the angular-space grid only in the **XZ** plane. This LID profile also resembles a near Lambertian such as that in Figure 7.15, but has much greater LID strength and distribution spread. (Data taken by Techno Team staff.)

the planes intact in these discussions by carefully studying the different intersecting planes along the **X**-, **Y**-, and **Z**-directions in Figure 7.17 because this 3-D graph shows a good amount of asymmetry in the LID as opposed to the nearly symmetric Lambertians in the other two figures.

The LEDGREEN pair shows a much wider distribution with larger LID strengths compared to the LFL-T8-S3/S4 and LED-T8-S1/S2 pairs. The total luminous flux obtained from the goniophotometric measurements for LEDGREEN, LEDTRONICS, and LFL-S3/S4 pairs was 4226, 2536, and 3740 lm respectively. The RiGO-801 LID and flux data support the illuminance data presented in Figures 7.10 and 7.11 in a comparative manner for the three lamp pairs. This is encouraging and meaningful because spatial flux and flux density quantities of a lamp are uniquely or unambiguously related to its luminous intensity distribution.

The measured data presented here show that the LEDGREEN-T8-S1/S2 pair generates a much higher luminous flux over wider regions compared to LED-T8-S1/S2 and LFL-T8-S3/S4 pairs for the same electrical input condition. The RiGO measurements yield luminous efficacies of 96 lm/W, 74 lm/W, and 58 lm/W for the LEDGREEN, LEDTRONICS, and LFL pairs respectively. It should be noted here that a number of parameters are not held constant in this

Figure 7.17. The measured LID data for LFL-T8-S3/S4 pair using a RiGO-801 system. The graph shows the lamp's LID in *XYZ* spatial coordinates in units of candela, while showing the angular-space grid only in the *XZ* plane. This LID profile on *XZ* planes in the y-direction spreads more widely compared to the profiles on *YZ* planes along the x-direction. (Data taken by Techno Team staff.)

comparison; these are the mismatches in CCT, power factor, and ballast or fixture geometric design, which all affect efficacy and flux distribution properties. However, these mismatches were not significant enough to alter the conclusion that the performances of the LEDGREEN **T8**s (LEDGREEN Company, New Jersey) are superior to the LFL and LEDTRONIC **T8**s in this study on several accounts.

Despite the lower efficacy seen in the preceding measurements, the advantage of LFLs is that their LID can be tailored to spread more widely with an appropriate luminaire or troffer design. In contrast, the LED T8s that place discrete SMT-LEDs on a tubular surface can only distribute higher flux levels over broader angles at the expense of an even higher degree of glare compared to the present case. As seen in Figure 7.17, the LID pattern from conventional LFLs shows a less concentrated and more uniform distribution over broader angular ranges, which can be further broadened with appropriate louver designs. This is illustrated in Figure 7.18, which shows the 2-D LID pattern from two industrial-grade LFLs similar to LFL-S1/S2 (see Figure 7.7, bottom fixture) and the LFL-S3/S4 samples described in Table 7.4, but presumably with a different troffer design.

The data can be found in the IES catalogue in LUMCat [150]. The graph shows normalized LID distributions in polar coordinates for two orthogonal planes. If one follows the *XYZ* coordinate system of Figure 7.15, the solid line in Figure 7.18

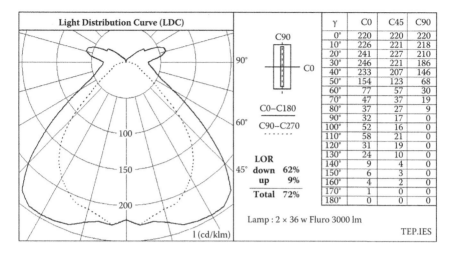

Light Distribution Curve (LDC)		γ	C0	C45	C90
		0°	220	220	220
		10°	226	221	218
		20°	241	227	210
		30°	246	221	186
		40°	233	207	146
		50°	154	123	68
		60°	77	57	30
		70°	47	37	19
		80°	37	27	9
		90°	32	17	0
		100°	52	16	0
		110°	58	21	0
		120°	31	19	0
		130°	24	10	0
		140°	9	4	0
		150°	6	3	0
		160°	4	2	0
		170°	1	0	0
		180°	0	0	0

C90

C0

C0–C180

C90–C270 ·······

LOR
down 62%
up 9%
Total 72%

90°

60°

45°

100

150

200

l (cd/klm)

Lamp : 2 × 36 w Fluro 3000 lm

TEP.IES

Figure 7.18. An IES archived LUMCat data presentation of light distribution in polar coordinates for an industrial-grade two-fluorescent-lamp luminaire, which has similar parameters as LFL-T8-S3/S4. The graph shows the LID profiles for C0–C180 (solid line) and C90–C270 (dotted line) planes.

would correspond to the LID in **XZ** plane at $y = 0$, and the dotted line would correspond to the LID in **YZ** plane at $x = 0$. For this analogy, one also needs to assume that the orientation of the luminaire is the same as that for the T8 lamps in Figures 7.15 through 7.17; this means that the lamp fixture is centered at the origin of the **XYZ** coordinate system with the LFL lengths aligned along the **Y**-axis. The wider spread in the **XZ** plane is expected because the larger lamp dimension (i.e., the LFL length) is parallel to the **Y**-axis.

In order to provide equivalency in the graphical representation, we now plot our LID data of LEDGREEN-T8-S1/S2 in polar coordinates in standardized units (candelas per kilolumens) in Figure 7.19, in the same manner the IES data are shown in Figure 7.18. This plot was generated using Techno Team's 3-D viewer.

Unquestionably, goniophotometric data of the kind shown in Figures 7.15 through 7.17 and Figure 7.19 provide the most revealing illumination characteristics of a lamp. It is the lamp's luminous intensity distribution in three-dimensional space, indicating both position and angular units simultaneously, which tells us both absolute and relative luminous flux strengths of the many points on the lamp surface by which the light source is characterized. Because such goniophotometric data can be complicated to obtain, those performing such tests are recommended to follow these procedures:

1. Choose angular resolutions appropriately for phi and theta based on the LID gradients; high resolutions are typically recommended for a more accurate LID distribution.

2. Perform burn-in and stability tests and record such data prior to LID measurements.

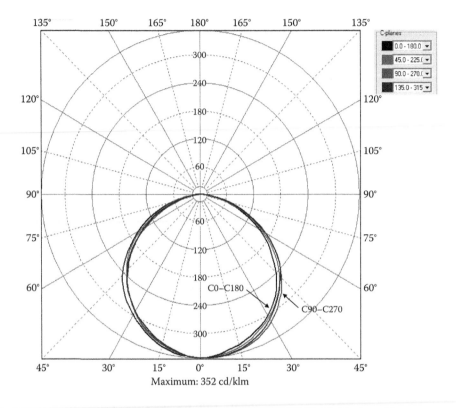

Figure 7.19. Various C-plane LID data of LEDGREEN-T8-S1/S2 plotted in a manner similar to the graph in Figure 7.18, which shows the LID data in two C-planes for a two-lamp LFL luminaire that has similar parameters as LFL-T8-S3/S4.

3. Set up apparatus according to the desired test conditions, such as temperature, current, voltage, etc.

4. Choose appropriate sample position and alignments corresponding to the customary usage of such lamps.

5. Assign appropriate directions for the sample and correlate them to the goniometer's *XYZ* coordinate system.

6. In order to comply with photometric principles that require the luminous intensity to be uniform over the chosen angular resolution, make sure that the measurement distance between source and detector (e.g., camera system) is at least 10 times the maximum dimension of the source.

The test engineers should be mindful that the goniometric measurement can be quite complex because of appropriate distancing and axial alignment accuracy of the camera system with respect to the test lamp sample. These types of goniophotometric data can only be reproduced by achieving the equivalent geometrical setup and lamp operating conditions as the prior case. If total flux is to be calculated from LID data, one must ensure that the lamp exhibits a uniform spatial flux distribution, void of any grossly irregular and unexpected patterns. Choice of angular resolutions may affect the chromatic data for LED lamps because their color characteristics typically exhibit rather pronounced angular dependence.

The measurement results presented in this section firmly establish that current LED replacement lamps are not yet ready to provide the same effective illumination we receive from linear fluorescent lamps to suit our commercial applications. In sum, this is primarily due to the arrangement of discrete LEDs in a planar array within a tubular structure that does not generate an equivalent luminous intensity distribution as that produced by a linear tubular fluorescent lamp with comparable structural dimensions. Such LED replacement lamps typically emit light with nongradual LID strengths over small and confined regions due to ineffective flux aggregation. In contrast, an LFL distributes light uniformly in all outward radial directions off the lamp surface, virtually all around the tube circumference and length, resulting in higher LID strengths over much broader regions. Therefore, large rooms or spaces can be more uniformly, brightly, and naturally illuminated with LFL arrays placed in the ceiling in clusters with a moderate spacing criterion. Because LED lamp technologies are improving fast and their luminous efficacy at small unit levels is already higher than that of fluorescent lamps, the SSL industry can now turn its focus on light distribution characteristics and illumination quality by inventing and adopting more advanced secondary optical as well as thermal designs.

7.6 Conclusion

Lighting has become a necessity in a very personal manner for many of our lives with hope to become more prevalent throughout the world. Its energy efficiency and quality are both vital factors. But, too often, consumers describe lighting quality only by the color properties, while many LED industry professionals describe it only by luminous efficacies. However, lumen distribution and appropriate brightness levels are also imperative to lighting quality, safety and achieving higher energy efficiency at the end-to-end system level.

Current LEDs' strengths make them suitable for signage, display, and directional lighting for proximity illumination applications because they are small, discrete, flat, bright, and mechanically durable. However, many of these characteristics also make them less desirable for high-quality, large-space, broad directional illumination applications. Individual LED lamps are rather small and very bright and, as efficacy improves, they will only get brighter at the unit level.

In contrast to organic LEDs, unit or discrete LEDs will remain flat for inorganic materials for the foreseeable future; tiling them to create large lamp surfaces is only practical when appreciable gaps exist between individual lamps, which are undesirable for creating uniform and broad illumination. The dark spots on the lamp surface generate uneven illuminance on planes of interests in many applications. In current typical LED lamp designs, increasing luminance from individual LED chips beyond a certain level in order to achieve a minimum required illuminance level throughout a large plane far away from the lamp does not provide a very desirable solution because it will simultaneously increase the luminance and illuminance properties beyond their maximum desirable levels, producing glare, wasting energy, and shortening LEDs' life span.

The technological, theoretical, and practical discussions in this book hopefully clarify that, while LED lamps have certain strengths, they clearly face challenges. Meanwhile, the incumbent lighting technologies also have several disadvantages; for example, incandescent lamps consume far too much energy and fluorescent lamps have weaknesses with respect to color properties and start-time and contain mercury.

Therefore, the need for new and improved lighting technologies is evident. The current environment is ready for a reform in lighting technologies that could, at best, create a constructive landscape by making designers and users conscious about energy savings while understanding and appreciating illumination quality. Numerous efforts around the world have begun to sprout with the goal of finding suitable lighting alternatives that would strike a proper balance between energy efficiency and aesthetics. The design concepts, simulation techniques, theoretical descriptions, measurement methods and results, and assessments presented in this book are intended to engage LED and lighting industry professionals, as well as academics, to develop beneficial LED lighting solutions further for consumers.

References

1. Sabra, A. I. 1982. *Theories of light from Descartes to Newton.* Cambridge: Cambridge University Press.
2. Huygens, C. 1690. *Traité de la lumiere,* ed. Pieter van der Aa, Chapter 1, Leiden, Netherlands. (Note: In the preface to his *Traité,* Huygens states that in 1678 he first communicated his book to the French Royal Academy of Sciences.)
3. Magie, W. F. 1935. *A source book in physics,* 309. Cambridge, MA: Harvard University Press.
4. Maxwell, J. C. 1865. A dynamical theory of the electromagnetic field. *Journal of Philosophical Transactions of the Royal Society of London* 155:459–512.
5. Kargh, H. December 2000. Max Planck: The reluctant revolutionary, *physicsworld*.com
6. Einstein, A. 1967. On a heuristic viewpoint concerning the production and transformation of light. In *The old quantum theory,* ed. Ter Haar, D. Pergamon. pp. 91–107. Added information (if needed): http://wien.cs.jhu.edu/AnnusMirabilis/AeReserveArticles/eins_lq.pdf (retrieved March 18, 2010).(The chapter is an English translation of Einstein's 1905 paper on the photoelectric effect.)
7. Griffiths, D. J. 2005. *Introduction to quantum mechanics,* 2nd ed. Upper Saddle River, NJ: Pearson, Prentice Hall.
8. Kane, R., and Sell, H. 2001. *Revolution in lamps: A chronicle of 50 years of progress,* 2nd ed., p. 37, Table 2-1. New York: Fairmont Press, Inc. p. 37, Table 2-1.
9. Sharpe, L. T., A. Stockman, A., Jagla, W., and Jägle, H. 2005. A luminous efficiency function, $V^*(\lambda)$, for daylight adaptation. *Journal of Vision* 5 (11): 948–968.
10. IESNA TM-11-00 (*Light trespass: Research, results and recommendations*): Illuminating Engineering Society of North America's (IESNA) document for outdoor lighting recommendations.

11. DiLaura, D., Houser, K., Mistrick, R., and Steffy, G. 2011. *The lighting handbook,* 10th ed. IES (Illuminating Engineering Society of North America).

12. US Environmental Protection Agency. 2012. Test methods: Wastes—hazardous waste. http://www.epa.gov/osw/hazard/testmethods (accessed December 23, 2012).

13. Vestel, L. B. April 9, 2009. The promise of a better light bulb, a blog about energy and the environment in *The New York Times.* http://green.blogs.nytimes.com/2009/04/09/the-promise-of-a-better-light-bulb/ (accessed December 23, 2012).

14. Khan, M. N. February 2011. LEDs marching towards general lighting. LED update column in *Signs of the Times.* Cincinnati, OH: ST Media Group International.

15. Lewin, I. 1999. Visibility factors in outdoor lighting design. *Proceedings of the 1999 Annual Conference of the Institution of Lighting Engineers.* The Institution of Lighting Engineers: United Kingdom.

16. Lewin, I. 2001. Lamp, color, visibility, safety and security. *Proceedings of Conference Seminar of Lightfair,* Las Vegas, May 30–June 1, 2001.

17. Silverstein, L. D. 2004. Color in electronic displays. *Society for Information Display Seminar Lecture Notes* 2, M-13/3-M-13/63.

18. Hunt, R. W. G. 2004. *The reproduction of color,* 6th ed. West Sussex, England: John Wiley & Sons Ltd.

19. Silverstein, L. D. 2006. Color display technology: From pixels to perception. *IS&T (The Society of Imaging Science and Technology),* vol. 21 (1), *The reporter (the window on imaging).*

20. *Encyclopædia Britannica.* 2009. 2006 Ultimate reference suite DVD: Eye, human. Also can be accessed online from Wikipedia: http://en.wikipedia.org/wiki/Encyclop%C3%A6dia_Britannica_2006_Ultimate_Reference_Suite_DVD (accessed December 23, 2012).

21. US Department of Energy. 2011. Solid-state lighting: Standards and development for solid-state lighting. http://www1.eere.energy.gov/buildings/ssl/standards.html (accessed December 23, 2012).

22. Braunstein, R. 1955. Radiative transitions in semiconductors. *Physical Review* 99 (6): 1892.

23. The quartz watch—Inventors. The first LEDs were infrared (invisible). The Lemelson Center. http://invention.smithsonian.org/centerpieces/quartz/inventors/biard.html (accessed January 5, 2013; retrieved August 13, 2007).

24. Nathan, M. I., W. P. Dumke, G. Burns, F. H. Dill, Jr., and G. Lasher. 1962. Stimulated emission of radiation from GaAs p-n junctions. *Applied Physics Letters* 1 (62). (Nathan's paper was received on October 6, 1962.)

25. Quist, T. M., R. H. Rediker, R. J. Keyes, W. E. Krag, B. Lax, A. L. McWhorter, and H. J. Zeiger. 1962. Semiconductor maser of GaAs. *Applied Physics Letters* 1 (91). (Quist's paper was received October 23, 1962, and in final form on November 5, 1962.)

26. Holonyak, N., Jr., and S. F. Bevacqua. 1962. Coherent (visible) light emission from $Ga(As_{1-x}P_x)$ junctions. *Applied Physics Letters* 1 (82). (Holonyak's paper was received on October 17, 1962.)

27. *The Japan Times Online.* Sept. 20, 2002. Court dismisses inventor's patent claim but will consider reward. http://www.japantimes.co.jp/text/nn20020920a2.html (assessed on January 4, 2013.)
28. Agrawal, G. P. 2002. *Fiber-optic communication systems,* 3rd ed. New York: John Wiley & Sons.
29. Sze, S. M. 1985. *Semiconductor devices—Physics and technology.* New York: John Wiley & Sons.
30. Wikipedia—The Free Encyclopedia. Millennium technology prize. http://en.wikipedia.org/wiki/Millennium_Technology_Prize (assessed on January 4, 2013).
31. Dai, Q., Q. Shan, J. Wang, S. Chhajed, J. Cho, E. F. Schubert, M. H. Crawford, D. D. Koleske, M. Kim, and Y. Park. 2010. Carrier recombination mechanisms and efficiency droop in GaInN/GaN light-emitting diodes. *Applied Physics Letters* 97:133507.
32. Kioupakis, E., P. Rinke, K. T. Delaney, and C. G. Van de Walle. 2011. Indirect Auger recombination as a cause of efficiency droop in nitride light-emitting diodes. *Applied Physics Letters* 98 (16).
33. Piprek, J. 2011. Unified model for the GaN LED efficiency droop. *Proceedings of SPIE 7939, Gallium Nitride Materials and Devices* VI:793916.
34. Flashlight news (press release by Osram Opto Semiconductors). 2010. Laboratory record: New chip platform increases LED efficiency by 30%. http://flashlightnews.org/story2978.shtml (accessed on January 5, 2013).
35. Iso, K., H. Yamada, H. Hirasawa, N. Fellows, M. Saito, K. Fujito, S. P. DenBaars, J. S. Speck, and S. Nakamura. 2007. High brightness blue InGaN/GaN light emitting diode on nonpolar m-plane bulk GaN substrate. *Japanese Journal of Applied Physics* 46 (40): L960–L962.
36. Overton, G. 2012. Bulk GaN substrate project from ARPA-E to be led by Soraa. *Laser Focus World.* http://www.laserfocusworld.com/articles/2012/08/arpa-e-bulk-gan-soraa.html (accessed on January 5, 2013).
37. Naoto, H., R.-J., Xie, and K. Sakuma. 2005. Science links Japan. New SiAION phosphors and white LEDs. *Oyo Butsuri* 74 (11): 1449–1452. http://sciencelinks.jp/j-east/article/200602/000020060205A1031052.php (accessed on January 5, 2013).
38. Montgomery, J. 2012. Solid state technology (insights for electronics manufacturing. Azzurro, Epistar achieve GaN-on-Si on 150 mm. http://www.electroiq.com/articles/sst/2012/10/azzurro-epistar-achieve-gan-on-si-on-150mm.html (accessed on January 5, 2013).
39. Whitaker, T. 2012. Osram Opto unveils R&D results from GaN LEDs grown on silicon. *LEDs Magazine.* http://ledsmagazine.com/news/9/1/19 (accessed on January 5, 2013).
40. Wong, W. S., T. Sands, and N. W. Cheung. 1998. Damage-free separation of GaN films from sapphire substrates. *Applied Physics Letters* 72 (5):599–601.
41. Paschotta, R. 2012. *RP photonics—Encyclopedia of laser physics and technology.* Light-emitting diodes. Last updated on March 21, 2012. http://www.rp-photonics.com/light_emitting_diodes.html (accessed on January 5, 2013).

42. Thurmond, C. D. 1975. The standard thermodynamic functions for the formation of electrons and holes in Ge, Si, GaAs, and GaP. *Journal of Electrochemical Society* 122:1133.

43. Kazempoor, M. 2009. Thermal management of sophisticated LED solutions. *LED Professional Review* May/June (13): 55.

44. Product News Category. 2012. GPD Global's PCD4H dispense pump improves yields for LED manufacturers. *LED Professional Review* 30:22.

45. Application Category. 2012. PCB design for a high end stage light. *LED Professional Review* 30:42.

46. Special Topics Category. 2011. High-brightness LEDs on FR4 laminates. *LED Professional Review* 27:62.

47. Product News Category. 2012. TE connectivity solderless LED socket for Nichia COB-L series LEDs. *LED Professional Review* 30:20.

48. Application Category. 2012. New approach for a modular LED COB system up to 500 W. *LED Professional Review* (30):38.

49. Neudeck, G. W. 1988. *The PN junction diode.* Upper Saddle River, NJ: Prentice Hall.

50. Rosen, R. 2011. Dimming techniques for switched-mode LED drivers. Texas Instruments (literature no.: SNVA605. http://www.ti.com/lit/an/snva605/snva605.pdf (accessed on January 7, 2013).

51. Product News Category. 2012. Fairchild LED driver for TRIAC-, analog- and nondimming lamp designs. *LED Professional Review* 30:14.

52. RECOM Lighting. DC input LED driver datasheets. http://www.recom-lighting.com/tools/datasheets.html (accessed on January 7, 2013).

53. Renesas News Release (Tokyo, Japan). December 10, 2010. http://www.renesas.com/press/news/2010/news20101210.jsp (accessed January 7, 2013).

54. Peters, L. 2012. Seoul semiconductor president Lee outlines AC-LED potential at SIL 2012. *LEDs Magazine* http://ledsmagazine.com/news/9/2/26 (accessed on January 7, 2013).

55. Kovach, L. D. 1983. *Boundary value problems,* 253. Reading, MA: Addison–Wesley Publishing Company.

56. *Sauna*™—Thermal modeling software from Thermal Solutions Inc. (Ann Arbor, Michigan). http://www.thermalsoftware.com/index.htm (accessed January 7, 2013).

57. Khan, M. N. 2010. How long do LEDs last? Part II: LED update column in *Signs of the Times.* Cincinnati, OH: ST Media Group International.

58. Street light brochure by EOI. SL2 LED street lights deliver excellence to roadway lighting, pp. 6–7. http://www.e-litestar.com/images/brochures/Brochure-SL2-long.pdf (accessed on January 8, 2013).

59. US DOE article, PNNL-SA-50957. 2009. Lifetime of white LEDs. http://apps1.eere.energy.gov/buildings/publications/pdfs/ssl/lifetime_white_leds.pdf (accessed on January 8, 2013).

60. IES—Illuminating Engineering Society 2008. Approved method: Measuring lumen maintenance of LED light sources. http://www.ies.org/store/product/approved-method-measuring-lumen-maintenance-of-led-light-sources-1096.cfm (accessed on January 8, 2013).

61. IES—Illuminating Engineering Society. 2011. Projecting long term lumen maintenance of LED light sources. http://www.ies.org/store/product/projecting-long-term-lumen-maintenance-of-led-light-sources-1253.cfm (accessed on January 8, 2013).

62. Khan, M. N. 2008. LED lighting technology fundamentals and measurement guidelines. *LED Professional Review* 10:14.

63. Nichia Corporation. Specification for white LEDs. Document no. NVSW219AT (cat. no. 110331). http://www.nichia.co.jp/specification/en/product/led/NVSW219A-E.pdf (accessed on January 8, 2013).

64. Schott. High brightness LED light line. http://www.us.schott.com/lightingimaging/english/machinevision/products/led-illumination/high-brightness-led-lightlines.html (accessed on January 8, 2013).

65. LED driver product specification from National Semiconductor (LM3433) (now Texas Instruments). http://www.digikey.com/product-search/en/integrated-circuits-ics/pmic-led-drivers/2556628?k=LM3433 (accessed on January 7, 2013).

66. Yoshida, T., Kawatani, A., and Shuke, K. SPIE digital library. Novel surface-mount type fiber optic transmitter and receiver. *SPIE Proceedings* http://proceedings.spiedigitallibrary.org/article.aspx?articleid=919377 (accessed on January 8, 2013).

67. Cox, A. 1946. *Optics: The technique of definition,* 6th ed. Waltham, MA: Focal Press.

68. Walker, J. G., P. C. Y. Chang, and K. I. Hopcraft. 2000. Visibility depth improvement in active polarization imaging in scattering media. *Applied Optics* 39 (27): 4933–4941.

69. Meyer-Arendt, J. R. 1968. Radiometry and photometry: Units and conversion factors. *Applied Optics* 7 (10): 2081.

70. Palmer, J. M., and B. G. Grant. 2010. *The art of radiometry.* Bellingham, WA: SPIE Press.

71. GL Optic Light measurement solutions. http://www.gloptic.com/products/ (accessed on January 9, 2013).

72. Gamma Scientific Light Measurement Solutions. http://www.gamma-sci.com/ (accessed on January 9, 2013).

73. Labsphere—A Halma Company. Spheres and components. http://www.labsphere.com/products/spheres-and-components/default.aspx (accessed on January 9, 2013).

74. Instrument Systems. Integrating spheres—luminous flux measurement for LEDs and lamps. http://www.instrumentsystems.com/products/integrating-spheres/ (accessed on January 9, 2013).

75. Konica Minolta Sensing Americas. Luminance meters. http://sensing.konicaminolta.us/technologies/luminance-meters/ (accessed on January 9, 2013).

76. Konica Minolta Sensing Americas. CL-500A illuminance spectrophotometer. http://sensing.konicaminolta.us/products/cl-500-illuminance-spectrophotometer/ (accessed on January 9, 2013).

77. Direct industry catalog search: Complete overview of RiGo goniopho-
    tometer—Techno Team Bildverarbeitung—#9; http://pdf.directindustry.
    com/pdf/technoteam-bildverarbeitung/complete-overview-of-rigo-gonio-
    photometer/63849-128609-_9.html (accessed on January 10, 2013).
78. Direct industry catalog search: LED station MAS 40 turn-key system for
    LED testing—Instrument systems—#4. http://pdf.directindustry.com/
    pdf/instrument-systems/led-station-mas-40-turn-key-system-for-led-
    testing/57082-68569-_4.html (accessed on January 10, 2013).
79. Bredemeier, K., R. Poschmann, and F. Schmidt. 2007. Development of
    luminous objects with measured ray data, Laser + Photonik, pp. 20–24.
    Web archived by Laser Photonics, EU: http://www.laser-photonics.eu/
    web/o_archiv.asp?ps=eLP100428&task=03&o_id=20080916104754-145
    (accessed on January 10, 2013). Direct link: http://www.laser-photonics.eu/
    eLP100428
80. XLENT Lighting Software. LUM Cat. http://xlentlightingsoftware.com/
    lighting-software/lumcat (accessed on January 11, 2013).
81. Schanda, J. 2007. *Colorimetry: Understanding the CIE system.* New York:
    John Wiley & Sons.
82. Wright, W. D. 1928. A re-determination of the trichromatic coefficients of
    the spectral colours. *Transactions of the Optical Society* 30 (4): 141–164.
83. Smith, T., and J. Guild. 1931–1932. The C.I.E. colorimetric standards and
    their use. *Transactions of the Optical Society* 33 (3): 73–134.
84. Fairman, H. S., M. H. Brill, and H. Hemmendinger. 1997. How the CIE
    1931 color-matching functions were derived from the Wright–Guild data.
    Color Research and Application 22 (1): 11–23.
85. Fairman, H. S., M. H. Brill, and H. Hemmendinger. 1998. Erratum: How the
    CIE 1931 color-matching functions were derived from the Wright–Guild
    data. *Color Research and Application* 23 (4): 259–259.
86. Davis, W., and Y. Ohno. 2010. Color quality scale. *Optical Engineering* 49
    (3): 033602.
87. CIE. 1995. Method of measuring and specifying colour rendering proper-
    ties of light sources. Publication 13.3, Vienna: Commission Internationale
    de l'Eclairage, ISBN 978-3-900734-57-2. (A verbatim republication of
    the 1974 second edition. Accompanying disk D008:Computer program to
    calculate CRIs.
88. Nickerson, D., and Jerome, C. W. 1965. Color rendering of light sources:
    CIE method of specification and its application, *Illuminating Engineering*
    (IESNA) 60 (4): 262–271.
89. Guo, X., and K. W. Houser. 2004. A review of color rendering indices
    and their application to commercial light sources. *Lighting Research and
    Technology* 36 (3): 183–199.
90. Bodrogi, P. 2004. Colour rendering: Past, present (2004), and future. *CIE
    Expert Symposium on LED Light Sources,* pp. 10–12. June 7–8, 2004,
    Tokyo, Japan.

91. Lighting issues in the 1980s. 1979. Summary and proceedings of a lighting roundtable held June 14 and 15, 1979, at the Sheraton Center, New York. Edited by A. I. Rubin, Center for Building Technology, National Engineering Laboratory, National Bureau of Standards, Washington, DC, 20234. (Sponsored in part by IESNA). http://www.getcited.org/pub/102115036 (accessed on January 10, 2013).

92. NIST—National Voluntary Laboratory Accreditation Program; NVLAP energy efficient lighting products LAP. http://www.nist.gov/nvlap/eel-lap. cfm (accessed on January 10, 2013).

93. Jaeggi, W. 2008. Tages Anzeiger. Grosses Lichterlöschen für die Glühbirnen (German). http://www.tagesanzeiger.ch/leben/rat-und-tipps/Grosses-Lichterlschen-fr-die-Glhbirnen/story/25999013 (accessed on January 12, 2013).

94. Australian Government. Department of Climate Change and Energy Efficiency. Lighting—Phase-out of inefficient incandescent light bulbs. http://www.climatechange.gov.au/en/what-you-need-to-know/lighting.aspx (accessed on January 12, 2013).

95. CBC News, British Columbia. 2011. Consumers hoard light bulbs amid B.C. ban. http://www.cbc.ca/news/canada/british-columbia/story/2011/01/25/consumer-incandescent-bulbs578.html (accessed on January 12, 2013).

96. Chandavarkar, P. 2009. Deutsche Welle. Kate Brown: Artists see EU light bulb ban as an aesthetic calamity. http://www.dw.de/artists-see-eu-light-bulb-ban-as-an-aesthetic-calamity/a-4594321-1 (accessed on January 12, 2013).

97. BBC. Switch off for traditional bulbs. http://news.bbc.co.uk/2/hi/uk_news/7016020.stm (accessed on January 12, 2013; last updated on September 27, 2007).

98. Porter, D. 2007. Edison's light bulb could be endangered. *USA Today.* http://usatoday30.usatoday.com/tech/news/2007-02-09-edison-bulb-ban_x. htm?csp=34

99. Associated Press. 2011. Trib live—USWorld. Congress flips dimmer switch on light bulb law. http://triblive.com/x/pittsburghtrib/news/s_772480. html#axzz2Hmiyyi2M (accessed on January 12, 2013).

100. Cardwell, D. 2010. When out to dinner, don't count the watts. *The New York Times,* N.Y/Region Section, written by Diane, June 7, 2010,http://www.nytimes. com/2010/06/08/nyregion/08bulb.html?_r=1 (accessed on January 12, 2013).

101. Blusseau, E., and L. Mottet. 1997. Complex shape headlamps: Eight years of experience (technical paper no. 970901 Society of Automotive Engineers). http://papers.sae.org/970901/ (accessed on January 12, 2013).

102. Sivak, M., T Sato, D. S. Battle, E. C. Traube, and Michael J. Flannagan. 1993. Mirlyn classic, legacy catalog of the University of Michigan Library. In-traffic evaluations of high-intensity discharge headlamps: Overall performance and color appearance of objects. University of Michigan Transportation Research Institute. http://mirlyn-classic.lib.umich.edu/F/?func=direct&doc_number=005512803&local_base=UMTRI_PUB (accessed on January 12, 2013).

103. Automotive lighting. LED in headlamps. http://www.al-lighting.com/lighting/headlamps/led/ (accessed on January 12, 2013).

104. LEDs Magazine. 2007. LED headlamp from Hella to appear on Cadillac. http://ledsmagazine.com/news/4/11/26 (accessed on January 12, 2013).

105. CALiPER Summary Report. 2008, Summary of results: Round 5 of product testing. http://apps1.eere.energy.gov/buildings/publications/pdfs/ssl/caliper_round_5_summary_final.pdf (accessed on January 13, 2013).

106. LED Professional. 2012. Osram offers DSL LED module to refurbish historic street luminaires. http://www.led-professional.com/products/led-modules-led-light-engines/osram-offers-dsl-led-module-to-refurbish-historic-street-luminaires (accessed on January 13, 2013).

107. The Climate Group. 2012. LED—Lighting the clean revolution. p. 23. http://thecleanrevolution.org/_assets/files/LED_report_web1.pdf (accessed on January 13, 2013).

108. Wright, M. 2012. Northeast Group research shows satisfaction with LED street lights. *LEDs Magazine* http://ledsmagazine.com/news/9/10/26 (accessed on January 13, 2013).

109. Zemax® is a registered trademark of Radiant Zemax LLC, Copyright 1990–2012. www.radiantzemax.com/en/zemax/ (accessed on January 14, 2013).

110. Mitsubishi Electric. Color TFT-LCD modules for industrial use: Super high brightness. http://www.mitsubishielectric.com/bu/tft_lcd/features/shbrightness.html (accessed on January 15, 2013).

111. Swokoski, E. W. 1981. *Calculus with analytic geometry,* 2nd ed., 929–931. Boston, Prindle, Weber, and Schmidt.

112. Cheng, D. K. 1985. *Field and wave electromagnetics,* Reading, MA: Addison–Wesley Publishing Company.

113. Leger, J. R., and G. M. Morris. 1993. Diffractive optics: An introduction to the feature. *Applied Optics* 32:14.

114. Feldman, M. R., W. H. Welch, R. D. Te Kolste, and J. E. Morris. 1996. *IEEE 46th Electronic Components and Technology Conference Proceedings,* 1278-1283.

115. Khajavikhan, M., and J. R. Leger. 2008. Efficient conversion of light from sparse laser arrays into single-lobed far-field using phase structures. *Optics Letters* 33:2377–2379.

116. Minano, J. C., P. Benitez, and A. Santamaria. 2009. Free-form optics for illumination. *Optical Review* 16 (2): 99–102.

117. Hoffman, A. 2011. Tailored optics for LED luminaires—From mass to low volume. LED Professional Symposium, Bregenz, Austria.

118. Wendel, S., J. Kurz, and C. Neumann. 2012. Optimizing nonimaging free-form optics using free-form deformation. *SPIE Proceedings,* vol. 8550, Barcelona, Spain.

119. Khan, M. N. 2013. Light distribution using tapered waveguides in LED-based tubular lamps as replacements for linear fluorescent lamps. US Patent No. 8348467, issued on January 8, 2013.

120. Khan, M. N. 2012. Patent pending. Date of origin: September 21, 2012.

121. Snyder, A. W., and J. D. Love. 1983. *Optical waveguide theory.* New York: Chapman & Hall.

122. Kaminow, I. P., and T. Li. 2002. *Optical fiber telecommunications,* vol. A, 4th edition: Components (optics and photonics). Waltham, MA: Academic Press.

123. Compound semiconductor. September 24, 2012. Remote phosphors yield better light bulbs. http://www.compoundsemiconductor.net/csc/features-details/19735527/Remote-phosphors-yield-better-light-bulb.html (accessed on January 15, 2013).

124. Burns, W. K. 1992. Shaping the digital switch. *IEEE Photonics Technology Letters* 4 (8): 861–883.

125. Khan, M. N., and R. H. Monnard. 2000. Adiabatic Y-branch modulator with negligible chirp. (Issued on May 16, 2000), US Patent 6064788.

126. Khan, M. N., B. I. Miller, E. C. Burrows, and C. A. Burrus. 1999. High-speed digital Y-branch switch/modulator with integrated passive tapers for fiber pigtailing. *Electronics Letters* 35 (11): 894–896.

127. Optiwave Software. OptiBPM by Optiwave Systems, Inc. http://www.optiwave.com/products/bpm_overview.html (accessed on January 15, 2013).

128. Paschotta, R. 2012. *Encyclopedia of laser physics and technology.* Heading: Multimode fibers; subheading: Multimode fibers for optical communications. RP Photonics. http://www.rp-photonics.com/multimode_fibers.html (accessed on January 15, 2013).

129. Koshel, R. J. 2013. *Illumination engineering: Design with nonimaging optics,* Chapter 6, Section 2. Hoboken, NJ: John Wiley & Sons.

130. Information Gatekeepers. 1993. *Plastic optical fiber design manual— Handbook and buyers guide.* Boston: Information Gatekeepers, Inc.

131. B. K. P. Horn. 1970. Shape from shading: A method for obtaining the shape of a smooth opaque object from one view. MIT Project MAC Internal Report TR-79 and MIT AI. Laboratory Technical Report 232.

132. Wald, M. L. 2012. Green—A blog about energy and the environment. A new bid for the 100-watt light bulb market, written. *The New York Times.* http://green.blogs.nytimes.com/2012/11/13/a-new-bid-for-the-100-watt-light-bulb-market/ (accessed on January 17, 2013).

133. Federal Trade Commission. 2013. Billing Code: 6750-01S, 16 CFR Part 305. Disclosures regarding energy consumption and water use of certain home appliances and other products under the Energy Policy and Conservation Act (Appliance Labeling Rule), p. 6, Footnote 11. http://www.ofr.gov/OFRUpload/OFRData/2013-00113_PI.pdf (accessed on January 17, 2013) or Federal Register/vol. 78 (6)/Wednesday, January 9, 2013/Proposed Rules. Page 1780, Footnote 11. http://www.gpo.gov/fdsys/pkg/FR-2013-01-09/pdf/2013-00113.pdf (accessed on January 17, 2013).

134. Hong, T., Kim, H., and Kwak, T. 2012. Energy-saving techniques for reducing CO_2 emissions in elementary schools. Journal of Management in Engineering 28 (1), Special issue: Engineering Management for Sustainable Development, 39–50. Web article in ASCE Library. http://ascelibrary.org/action/showAbstract?page=39&volume=28&issue=1&journalCode=jmenea (accessed on January 17, 2013).

135. IEA (International Energy Agency). 2010. Annex 45, energy efficient electric lighting for buildings. http://www.lightinglab.fi/IEAAnnex45/ (accessed on January 18, 2013).

136. EIA. 2000. Commercial office buildings—How do they use electricity? Release date: September 11; last modified: January 3, 2001. http://www.eia.doe.gov/emeu/consumptionbriefs/cbecs/pbawebsite/office/office_howuseelec.htm) (accessed on January 18, 2013).

137. Osram. 2005. ECG for T5 fluorescent lamps—Technical guideline, p. 6. http://www.osram.es/_global/pdf/Professional/ECG_%26_LMS/ECG_for_FL_and_CFL/130T015GB.pdf (accessed on January 18, 2013).

138. IEA. 2010. Annex 45 guidebook. Chapter 5: Lighting technologies, table 5-2, p. 97. http://lightinglab.fi/IEAAnnex45/guidebook/5_lighting%20technologies.pdf (accessed on January 18, 2013).

139. LUXADD, Express T5 retrofit kit for T12 with magnetic ballast, copyright 2010–2012 LUXADD LLC. http://www.luxadd.com/index.php/luxadd-double-lamp-express-retrofit-kit-t12-t5.html (accessed on January 18, 2013).

140. US Department of Energy (DOE), Office of Energy Efficiency and Renewable Energy, Federal Energy Management Program. 2000. How to buy energy-efficient fluorescent ballasts. 2000. http://www1.eere.energy.gov/femp/pdfs/ballast.pdf (accessed on January 18, 2013).

141. IES (Illuminating Engineering Society). 2011. Technical memorandum: IES TM-23-11. Lighting control protocols.

142. USITT (United States Institute for Theatre Technology). 2012. DMX512 FAQ. http://www.usitt.org/content.asp?contentid=373 (accessed on January 18, 2013).

143. Tridonic. DSI interface—luxCONTROL lighting control systems. http://www.tridonic.com/com/en/products/386.asp (accessed on January 18, 2013).

144. CALiPER Summary Report. October 2010. Round 11 of product testing. US Department of Energy, p. 3. http://apps1.eere.energy.gov/buildings/publications/pdfs/ssl/caliper_round-11_summary.pdf

145. CALiPER Summary Report June 2011. Round 12 of product testing. US Department of Energy. http://apps1.eere.energy.gov/buildings/publications/pdfs/ssl/caliper_round12_summary.pdf (accessed on January 18, 2013).

146. Khan, M. N. February 2009. Understanding energy efficiency. LED/EDS column in *Signs of the Times*. Cincinnati: ST Media Group International.

147. Irujo, T. 2011. Optical fiber in enterprise applications; OM4—The next generation of multimode fiber. OFS—A Furukawa Company. http://www.ofsoptics.com/resources/OM4-The-Next-Generation-of-MMF.pdf (accessed on January 18, 2013).

148. Young, G. 2010. Scenic America, sign brightness, measuring sign brightness. http://www.scenic.org/storage/documents/EXCERPT_Measuring_Sign_Brightness.pdf (accessed on January 18, 2013).

149. CALiPER Summary Report. Round 11 of product testing. US Department of Energy, pp. 29–30.

150. LUMCat, version 3.5. Copyright 1999/2002 by Xlent. Distributed by Crossman Consulting, Australia.

Index

Printed and bound by CPI Group (UK) Ltd, Croydon, CR0 4YY

01/11/2024

01782625-0004